Institutional and Welfare Aspects of the Provision and Use of Information and Communication Technologies in the Rural Areas of Bangladesh and Peru

DEVELOPMENT ECONOMICS AND POLICY

Series edited by Franz Heidhues and Joachim von Braun

Vol. 30

PETER LANG
Frankfurt am Main · Berlin · Bern · Bruxelles · New York · Oxford · Wien

Institutional and Welfare Aspects of the Provision and Use of Information and Communication Technologies in the Rural Areas of Bangladesh and Peru

Shyamal K. Chowdhury

PETER LANG
Europäischer Verlag der Wissenschaften

Die Deutsche Bibliothek - CIP-Einheitsaufnahme

Chowdhury, Shyamal K.:

Institutional and welfare aspects of the provision and use of information and communication technologies in the rural areas of Bangladesh and Peru / Shyamal K. Chowdhury. - Frankfurt am Main ; Berlin ; Bern ; Bruxelles ; New York ; Oxford ; Wien : Lang, 2002
(Development economics and policy ; Vol. 30)
Zugl.: Bonn, Univ., Diss., 2002
ISBN 3-631-50161-7

Printed with the support of the
German Academic Exchange Service (DAAD).

D 98
ISSN 0948-1338
ISBN 3-631-50161-7
US-ISBN 0-8204-6044-3

Printed in Germany 1 2 4 5 6 7

www.peterlang.de

Dedicated to the memory of my father, the late Sharadindu Roy Chowdhury

Foreword

Access to and the use of information and communication technologies (ICTs) have emerged as one of the most important area of policy debate in recent time.

After volume 26, 27 and 29 of *Development Economics and Policy* on 'Telecommunication Services in Sub-Saharan Africa' by Romeo Bertolini, 'Use and Impact of Information and Communication Technologies in Developing Countries' Small Businesses: Evidence from Indian Small Scale Industry' by Dietrich Müller-Falcke, and 'The Impact of Information and Communication Technologies on Farm Households in China' by Wensheng Wang, this volume has focused on Bangladesh and Peru. It has addressed four research questions: first, why access to telecommunications is important; second, how to provide access to telecommunications for the people living in rural areas of developing countries; third, what are the direct welfare implications of the use of telecommunications; and fourth, what are the indirect welfare implications of the use of telecommunications. The volume has addressed these four issues under the institutional aspects of the provision of telecommunications services and the welfare aspects of the use of telecommunications services. The empirical findings of the study are based on household surveys that were conducted in the rural areas of the two study countries, Bangladesh and Peru.

The findings of the study reveal that there are both economic and development rationales that call for the provision of universal service. The two alternative institutional mechanisms to achieve universal services discussed in this study are public-private partnership in Peru and business-NGO partnership in Bangladesh are two examples of the use of the market in achieving universal access for people living in rural areas of these two countries.

The study has measured the direct welfare implications of the use of telecommunications through the measurement of Hicksian compensating variation (CV) and equivalent variation. The estimated CVs show that for all four income-quartiles, households derive a positive CVs stemming from the availability of telecommunications in rural areas in Bangladesh and Peru. The measurement of equivalent variation through parametric and non-parametric estimations suggests that rural telecommunications projects are welfare enhancing, since households' willingness to pay are higher than the prevailing tariff rates.

In case of indirect welfare impacts, the parametric estimation that controlled for other observable factors finds that telephone use has a sizable impact on households' factor market participation. In addition, the study has also revealed that access to information in the form of access to telephone has significant impact on choice of marketing channels and of selling intensity of rural producers. The lower the information cost, the higher the probability of selling to direct buyers and higher the selling intensity to direct buyers.

It is hoped that the present study reduces the information gaps that persist between the researchers and the policy makers on the importance of universal access, the benefits of the access and the institutional mechanisms to provide access of the rural people of developing countries. It is also hoped that the study reduces the information asymmetry of business communities that usually consider rural households are not willing to pay sufficient enough to extend ICTs.

Prof. Dr. Joachim von Braun
Center for Development Research (ZEF)
University of Bonn

Prof. Dr. Franz Heidhues
Center for Tropical Agriculture
University of Hohenheim

Table of Contents

List of Figures, Tables, Boxes and Maps

Figures

Tables

List of Tables in the Appendixes

Boxes

Maps

Acronyms and Abbreviations

AMPS	Advanced Mobile Phone System
BRTA	Bangladesh Rural Telecom Authority
BTTB	Bangladesh Telegraph and Telephone Board
CDMA	Code Division Multiple Access
COLR	Carrier of Last Resort
CPT	Compañia Peruana de Teléfonos
CS	Consumer Surplus
CV	Compensating Variation
ENTEL	Empresa Nacional de Telecomunicaciones
EV	Equivalent Variation
FITEL	Fondo de Inversión en Telecomunicaciones
FTC	Fixed Transaction Cost
GB	Grameen Bank
GDP	Gross Domestic Product
GOB	Government of Bangladesh
GP	GrameenPhone
GRADE	Grupo de Análisis para el Desarrollo
GSM	Global System for Mobile
GSM	Global System for Mobile Communications
GTC	Grameen Telecom
ICT	Information and Communication Technology
ILD	International Long Distance
ISP	Internet Service Providers
ITU	International Telecommunication Union
JU	Jahangirnagar University
LCU	Local Currency Unit
MC	Marginal Cost
MOPT	Ministry of Posts and Telecommunications
MTC	Ministry of Transportation and Communications
NGO	Non Government Organization
NLD	International Long Distance
OSIPTEL	Organismo Supervisor de la Inversión Privada en Telecomunicaciones
PC	Personal Computer
PCO	Public Call Office
PPP	Purchasing Power Parity
PTC	Proportional Transaction Cost
SMS	Short Message Service
TAI	Technology Achievement Index
TC	Transaction Cost
TCE	Transaction Cost Economics

TCS	Total Consumer Surplus
TDMA	Time Disvision Multiple Access
TDP	Telefónica del Perú S.A
TRB	Telecommunication Regulatory Board
USO	Universal Service Obligation
UUCP	Unix-to-Unix Copy
VMS	Voice Mail Service
VPP	Village Pay Phone
VSAT	Very Small Aperture Terminal
WTP	Willingness to Pay
ZEF	Zentrum für Entwicklungsforschung

Acknowledgements

I would like to extend my sincere gratitude and appreciation to my supervisor Prof. Dr. Joachim von Braun for his ideas, inspirations and continuous guidance during my study at the Center for Development Research (ZEF). Without his guidance, this work would not have been possible in its present form. Other main debt is the intellectual influence of Dr. Maximo Torero for his numerous comments and suggestions that helped significantly to improve the quality of the work. I would also like to thank Prof. Dr. Gerhard Schiefer for his willingness to become my co-supervisor and for his valuable comments.

I have been benefited from colleagues at ZEF, particularly from Dr. Susanna Wolf, Dr. Ulrike Grote, Dr. Maja Micevska, Alexandra Schleier, Daniela Lohlein, Karin Astrid Siegmann and members of the ICT group, particularly from Romeo Bertolini, Dietrich Müller-Falcke, and Gi-Soon Song, and I would like to thank all of them. Finally, I would like to thank all my friends and colleagues those who helped me through encouragements and moral supports.

1 INTRODUCTION

1.1 Background

Is access to information and communication technologies important? If access is important, how to provide it for those people who live in rural areas in developing countries? Do rural households demand telephones? Why do some farm households participate in markets and others do not? Does access to information in the form of access to telecommunications bring any change in the farm households' behaviour? The present dissertation examines these questions under two areas: i) institutional aspects and ii) welfare aspects of information and communication technologies in the context of rural areas of two countries, Bangladesh and Peru. The two study aspects, institutional and welfare, can also be viewed as supply of and demand for information and communication technologies respectively. Under the institutional aspects, the present study examines economic and development rationales behind the provision of information and communication technologies and alternative provision mechanisms. Under the welfare aspects, the study examines the demand for information and communication technologies (ICTs), and the direct and indirect welfare impacts of such technologies on the behaviour of a household. In the study, telecommunications and ICTs are used synonymously. Though ICTs include a wide range of services, due to their limited availability in the rural areas of the two study countries, the present study focuses on telecommunications services only.

Information and communication technologies include two technologies, information and communications, defined as 'electronic means of capturing, processing, storing, and communicating information'[1], and 'involve innovations in microelectronics, computing (hardware and software), telecommunications and opto-electronics – microprocessors, semiconductors, fibre optics'.[2] Rapid innovations in information and communication technology sectors are turning these technologies 'faster, better, cheaper'.[3]

Technological innovation affects human development in two ways: first, it increases the human capabilities that ensure more active participation in economic and social life; second, technological innovation increases the productivity. In the last century, gain in productivity that was the primary factor behind the real income growth came mainly from technological change. In fact the general advancement in human development and poverty reduction in the 20th century came largely from the technological breakthroughs.[4] Information

---

[1] Heeks (1999).

[2] UNDP (2001), p. 32.

[3] Jorgenson (2001), p. 2.

[4] UNDP (2001), pp. 2, 28-29; see also Romer (1986, 1990)

and communication technologies are considered as general-purpose technologies with pervasive application potentiality.[5]

Information and communication technologies, henceforth ICTs, have received enormous attention both in the context of industrialized and developing countries in recent times. Views on the effects of ICTs on developing countries range from wild optimism to deep pessimism.[6] Some observe that information and communication technologies are contributing 'to revolutionary changes in everyday life'[7], and 'it is only a matter of time before market mechanisms ensure that all the world's citizen enjoy the benefits of global information society'.[8] Opponents of such views argue that 'the transformative nature of these technologies has been exaggerated.[9] The new ICT applications may destroy more jobs than they create; the gap between rich and poor may widen; and the huge capital investments required to strengthen national capabilities for using ICTs could divert resources from other activities that could have greater development impact'.[10] However, technology itself is not good or bad per se and effects of any technology depend mostly on its use.

Despite the lack of general agreement on the impact of ICTs, the decade of the 1990s has seen a significant increase in investment in ICT infrastructure. During this period, annual investment in ICTs has increased more than two fold from about US$80 billion in the early '90s to about US$190 billion in the late '90s, of which about US$95 billion was in developing and transition economies.[11] This investment in ICTs represents around 12% of total GDP of low-income countries. As ICTs are characterized by falling in price and improvements in quality, the actual value of quality-adjusted investment could be higher. Table-1.1 shows the recent expansion in global information and communication infrastructures. Ever since, more and more people are getting access and demanding ICTs. For instance, the total number of Internet users has increased from fewer than 20 million in 1995 to more than 400 million in 2000. During the same period, the total number of websites has increased from fewer than 200 websites in mid-1993 to 20 million websites in late 2000.[12]

---

[5] OECD (1988) recognizes information technologies as the most pervasive technical innovation of the post-war era.

[6] See Morales-Gomez and Melesse (1998) for a positive view and Avgerou (1998) for a more skeptical view. See also Bedi (1999) for a review.

[7] See Mansell (1999).

[8] See Tapscott (1995)

[9] See Kubicek et al (1997); see also Sussman (1997).

[10] See Mansell (1999) p. 36.

[11] Wellenius et al (2000).

[12] UNDP (2001), p. 32.

Table-1.1: Expansion in Selected ICT Infrastructures: World, Bangladesh and Peru (in million units)

Indicators	1990	1999	Growth Rate (%)
World			
Fixed Telephone	480	890	7
Mobile Telephone	10	430	52
Internet Account	7	200~	40
Website (1993-2000)	0.0002~	20~	418
Bangladesh			
Fixed Telephone	0.24	0.43	9
Mobile Telephone (1992-99)	0.00025	0.15	8500
Internet Account (1997-99)	0.001	0.05	2450
Peru			
Fixed Telephone	0.57	1.63	21
Mobile Telephone	0.00165	1.01	6813
Internet Account (1995-99)	0.008	0.4	1225

Source: Wellenius et al (2000); ITU (2001); ~ data from UNDP (2001); for Internet, the last data point is for the year 2000; n.a.: not available.

Similar to the worldwide trend, the two countries on which the present study concentrates, Bangladesh and Peru, have also shown phenomenal expansion in selected ICT infrastructure in recent times. In fact both Bangladesh and Peru have achieved a higher growth in fixed telephony, mobile telephony and Internet users in comparison to the World as a whole.

Notwithstanding these remarkable advancements in information and communication technologies in recent times, a 'digital divide' exists between industrialized and developing nations and between urban-rich and rural-poor within nations. Table-1.2 provides some of the indicators pertaining to ICTs showing the state of access between countries and within a country. More than three-quarters (79 percent in 2000) of the Internet users live in high-income OECD countries that comprise less than a quarter (14 percent in 2000) of the world's population. Within the developing countries, users are predominantly urban-rich educated males.[13] This implies that access is limited mostly to a sub-group of countries and to a sub-group of people. So the question remains as to how to ensure access for developing countries to ICTs and, within developing countries, how to ensure access of rural poor people to ICTs.

[13] UNDP (2001), p. 40.

Table-1.2: Inequality in Access to ICTs: between Countries and within a Country (1999)

Indicators	Low Income~	High Income~	World	Bangladesh	Peru
Information					
Personal Computer§	0.71	34.80	6.84	0.10	3.57
Communication					
Main telephone lines§	4.32	58.50	15.16	0.34	6.45
Cellular telephone §	1.38	37.79	8.22	0.11	4.01
Internet users§	0.38	20.88	3.98	0.39	1.59
*Ratio of urban-rural teledensity**	2.56	1.17	2.73	44.49	47.5

~World Bank's definition; § Figures are for per 100 inhabitants; *Ratio of largest city teledensity to rest of the country. Source: ITU (2001) and World Bank (2000).

Turning to the study countries, both countries position below the high-income countries and below the world average. However, the digital divide is more prominent in the case of Bangladesh than in the case of Peru. As a country, Bangladesh positions below the World average and below the low-income countries. In the case of the digital divide within a country, both Bangladesh and Peru show substantial digital divide between rural areas and urban areas manifested by the highly skewed distribution of telephone between the rural areas and urban areas.

Against the above background, this study examines the institutional and welfare aspects of rural-poor people in developing countries to ICTs. More specifically, it draws experiences from rural areas of two developing countries, Peru and Bangladesh, and conducts an empirical examination of both access and welfare aspects of ICTs. Though information and communication technologies include a wide range of services, the present study focuses on telecommunication services for two reasons. First, telecommunication services are viewed as precursors of other advanced ICT services, and second, the extent of other form of ICT services are either very limited or do not exist in the rural areas of Bangladesh and Peru.

1.2 Research Questions

Due to the two distinct focuses of the present study, all the research questions have been arranged under two parts:

i. Institutional aspects of the provision of ICTs
ii. Welfare aspects of the use of ICTs

The first part, institutional aspects, deals with the supply of ICTs, and the second part, welfare aspects, looks at the benefits of the use of ICTs. In other words, while the first part examines the institutional provision mechanisms, the second part examines the benefits of such provision mechanisms through the measurement of welfare. In order to address the specific research questions, telecommunications is assumed to be a direct proxy for ICTs. Accordingly, the term telecommunications is used as a synonym for ICTs.

1.2.1 Institutional Aspects of the Provision of ICTs

This part tries to answer the following two questions:

i. Why access to telecommunications is important?
ii. How to provide it?

The present study will start with the research question 'why access to telecommunications is important?' More specifically, it will ask: what are the economic and development rationales to provide access to telecommunications? The study will discuss this question under the general concept of universal service obligation (USO). In addition, it will also discuss this issue under two different institutions, state and private markets, and will provide experiences of Bangladesh and Peru in achieving universal service under these two institutional mechanisms.

The second research question that the study will deal with is 'how to provide access to telecommunications services for people living in rural areas in developing countries'. It will develop on the theory described under the previous research question and will ask: If access to telecommunications is important, how to provide it? How to extend telecommunications, particularly to rural areas in developing countries? What are the new promises in this regard?

To answer this set of questions, the study will review the traditional institutions, i.e. telecommunications provision under state and market mechanisms, and will evaluate two best practices, business-NGO partnership in Bangladesh and public-private partnership in Peru, that provide access for rural people to telecommunications.

1.2.2 Welfare Aspects of the Use of ICTs

Once the 'why' and 'how' questions are answered in the first part, it is necessary to examine the actual benefits that the households in rural areas receive from the use of telecommunications provided through the best practices described under the institutional aspects. For this, in the second part the study will answer two questions related to welfare:

i. What are the direct welfare impacts of telecommunications on rural households?

ii. What are the indirect welfare impacts of telecommunications on rural households?

To answer the two research questions related to welfare aspects, the study will sketch a simple framework where both direct and indirect welfare aspects will be discussed. To measure the direct welfare impacts, the study will rely on Hicksian compensating variation and equivalent variation and will measure a rural household's willingness to pay.

The fourth research question that the study will answer is 'what are the indirect welfare impacts of the use of telecommunications'. To put it another way, does access to telecommunications bring any change in a household's economic behaviour, particularly in transaction behaviour? Can it help to overcome the market failures in subsistence agricultural societies?

The current research will answer this question under the transaction costs framework where transaction costs will be viewed as the cause of market failure. Considering information cost as a part of transaction costs, the study will examine whether access to information is sufficient enough to bring discrete change in a household's market participation behaviour. The study will examine a household's participation in the case of both product and factor markets.

1.3 Why Bangladesh and Peru?

The empirical analysis of the present study will be based on data from two countries, Bangladesh and Peru. There are both similarities and differences between Bangladesh and Peru that should make the comparison an interesting one. As the review of the ICT sector will reveal, in terms of similarities both Bangladesh and Peru are characterized by under provision of telephone networks for households living in rural areas despite the fact that both countries have a competitive rural telecom market. Coincidentally, both Bangladesh and Peru currently contain two examples that are considered as two best practices in rural telecommunications provision.[14] In terms of differences, one can identify three major ones. First, though recent developments in both countries are enabling rural communities to have limited access to telecommunications services, differences exist in the provision mechanism. While in Peru it is through regulation and private sector participation, in Bangladesh a business non-profit partnership is providing telecommunications services to people living in rural areas. The second difference is in terms of technology. While in Peru it is fixed telephony, in Bangladesh it is mobile phones. The third difference comes from the difference in geography. While with the exception of the eastern part Bangladesh has a flat terrain, Peru is characterized by the

[14] See Bayes et al (1999) for Bangladesh and Cannok (2001) for Peru.

presence of mountains and difficult terrains that make the establishment of any network a complicated one.

1.4 Data Collection and Survey Methods

The present study will use data from two household surveys conducted in the six different rural regions in Bangladesh and the south rural region of Peru. For both countries, surveys have followed a two-stage random sampling procedure in selecting households and contained questions on households' demographic and housing characteristics, employment, income and expenditure, participation in social programs, and availability of infrastructure, access, use and willingness to pay for telephones, among other things. Both surveys have included settlements with and without public telephones where a lack of a public telephone service was primarily due to a supply constraint instead of a lack of demand.[15] Appendix-I provides the details of the survey designs for Bangladesh and Peru.

1.5 Structure of the Study

The study is structured in two parts: part one deals with the institutional aspects of the provision of ICTs and part two deals with the welfare aspects of the use of ICTs. In the first part, the unit of analysis is institution and, in the second part, the unit of analysis is rural household. However, in the first part as well, the analysis of institutions is centred on households.

The first part of the study is comprised of three chapters. Chapter 2 draws on the economic and development rationales of universal service obligation and answers the question 'why access to telecommunications is important. Chapter 3 is descriptive in nature; it reviews the state of the market and regulatory structure of the information and communications markets in the study countries. It also reviews the state of universal service in telecommunications under the traditional institutions: state, regulated monopoly and competition. Chapter 4 is based on the previous two chapters; it answers the question of how to ensure universal service for households living in rural areas of developing countries. Under the circumstances when traditional institutions do not voluntarily provide universal service to this group of households, it evaluates two alternative provision mechanisms, public-private partnership in Peru and business-NGO partnership in Bangladesh in universal service provision.

The second part of the study is comprised of five chapters and deals with the welfare implications of the use of information and communication technologies, particularly telecommunications. This part starts with chapter 5, which develops a conceptual framework that is utilized as a guiding principle in analysing the welfare implications of telecommunications in subsequent chapters. Based on

---

[15] Many of the settlement inhabitants without access to public telephone services travel many kilometers to arrive at a settlement that has the service.

this, chapter 6 and chapter 7 measure the direct welfare implications of use of telecommunications through the measurement of compensating variation and equivalent variation respectively. The next two chapters deal with the indirect welfare implications of use of telecommunications, and examine rural households' product and factor market participation behaviour associated with the use of telecommunications.

The study ends with a summary and conclusions described in chapter 10.

Part-I: Institutional Aspects of the Provision of ICTs

2 ACCESS TO ICTS AND UNIVERSAL SERVICE OBLIGATION

2.1 Introduction

This chapter and the subsequent two chapters deal with the institutional aspects, i.e. the access to ICTs. To ensure access, the concept that has been developed over time, especially in the case of telecommunication services, is universal service obligation. The phrase 'universal service' was first brought into use in 1907 by Theodore Vail, the then President of AT&T, to advocate that Bell should become the sole US telephone service provider to achieve widespread geographic access. Since then, the concept of 'universal service' has undergone a great deal of revision, and there have been constant concerns to define the term, and Vail's 'One System, One Policy, Universal Service' slogan has little to do with the contemporary notion of universal service.[16] However, despite continuous changes, it is possible to trace some common factors that determine the universal service bundle.

In the case of telecommunications, the primary objective of universal service in developing countries is to 'break the isolation for communities and households living in rural and remote areas' by bringing the households into a basic communication network. The primary benefit of such service is to reduce the cost of communications for the communities and for outsiders which might then result in changes in production and marketing structures in the communities. Apart from an increase in commercial activities, bringing households to the network can also reduce the delivery costs of other public services such as education and health.

This chapter will answer the research question, 'why is access to telecommunications important?'. For this, it will review the concept of universal service obligation and its importance. It will also briefly review 'universal access' – a concept closely related to universal service but which has more applicability in the context of rural areas of the developing countries. It will examine the issue of universal service under different market structures. The review will conclude with examining different universal service practices.

2.2 What is Universal Service Obligation?

A universal service obligation, henceforth USO, is typically viewed as the obligation of an operator to provide to all users a 'basic' package of services of 'good' quality and at 'affordable' prices.[17] In general, there are at least three restrictions that USO imposes on a service provider. These are quality, widespread availability, and uniform pricing. These three constraints are interrelated and need to be met simultaneously. Quality as an explicit restriction ensures the operator's minimum level of 'effort' to avoid moral hazard

---

[16] Cited in Blackman (1995)

[17] See Valletti (2000).

problems, and availability ensures that every household has the option to subscribe to the telecommunications network. The uniform pricing restriction is a binding constraint and ensures non-discriminatory access of all users to telecommunications.

2.3 Universal Service versus Universal Access

A notionally similar but more applicable concept of universal service in the context of developing countries is universal access. In contrast to industrialized countries that define the access to telecommunications at household level, the notion of universal access seeks to ensure access at a 'reasonable' level, e.g. at community level. While the maintenance of universal service in telecommunications is a cornerstone of public policy in industrialized countries, its minimum achievement in rural areas has remained the key public policy in developing countries.

The present study focuses on universal access, though the criteria on which universal access will be evaluated are based on universal service principles. Though one can draw a distinction between the two and universal access is often viewed as a precursor of universal service, universal service and universal access are closely related to each other. Moreover, though the distinction can have applicability in the context of industrialized countries, such a distinction plays only a minor role for the present purpose.[18]

2.4 Economic and Development Rationale Behind Universal Service

Though there are both political and economic rationales behind USO, and the obligation that any regulatory authority imposes on a service provider under universal service regulation is determined by both economic and political aspects, the present study keeps the discussion limited to the economic rationale only. In the case of telecommunications, one can find as many as four main reasons for imposing universal service obligations: income redistribution, the presence of 'merit goods', regional policy and network externalities.[19] However, two prime rationales that are widely cited in economic literature are network externalities and income redistribution.[20]

2.4.1 Network Externalities

Any industry is said to have network externality when the benefit to a consumer from the network increases with the number of other consumers or with the size of the network.[21] Telecommunications networks are characterized by the presence of positive network externalities as each additional subscriber in the network increases the utility of existing subscribers without imposing any cost

[18] See Federal Communications Commission (1999) for detailed.
[19] See Crémer (2000).
[20] See Panzar (2000).
[21] See for example Haruvy and Prasad (1998) on software.

on them. The presence of network externality can result in an inefficient outcome in an unregulated market. This is because the individual's subscription decision depends on a comparison between his/her own utility and the subscription price though the total utility that such a subscription generates in the network could be higher than the subscription price. As the individual subscriber does not take into account this externality, the size of the network could remain too small under market provision.

Until recently, the public provision of telecommunications has been justified on the grounds of the public good as well as natural monopolies. The dominant aspect of the public good is the network externalities that telecommunications offer. The natural monopoly in telecommunications provision arises essentially from the 'network' that is necessary to 'transport' the service, which provides common transportation facilities for all users rather than being dedicated to individual ones.[22]

2.4.2 Redistribution Mechanism

One important rationale behind the USO is the indirect redistribution mechanism. Under uniform price restriction, USO ensures price equality between high-cost and low-cost areas. Through subsidizing the 'basic package' or 'connection charge' by charging higher prices for other services, USO can be used in favour of low-income groups who otherwise would not subscribe to the network. Such a policy can ensure redistribution towards high-cost areas and low-income groups, and increases network externalities as well.

The role of USO as a redistributive mechanism can also be viewed as a development rationale. In fact, some economists find this the most compelling theoretical justification.[23] In situations where standard redistributive instruments suffer from problems of information asymmetry and are politically costly to implement, regulators can utilize USO as a redistributive mechanism.

2.4.3 Merit Good and Regional Policy Instrument

The need for universal service in telecommunications can be justified since a telecommunications network has merit good characteristics. USO can also be a regional policy instrument as it can be used to achieve specific development objectives for a region.

Telecommunications networks possess attributes of infrastructure and they are essential to modern trade, and indeed to modern living.[24] As infrastructure, telecommunications act as intermediaries, influence the marginal productivity

---

[22] See Kay (1994) on this monopoly argument related to common network.

[23] See for example, Cremer et al (1998).

[24] Crandall (1997).

of private capital, and allow the achievement of specific policy objectives such as spatial diversification.

In addition, a telecommunications infrastructure has a crowding-in effect. An adequate telecommunications infrastructure can act as a complementary capital to private capital and could be growth enhancing. Though an individual firm can partly substitute for deficient infrastructure through private investment, infrastructure like telecommunications that creates network externalities can hardly be substituted.

Telecommunications infrastructure can also be used as a policy tool to achieve specific development objectives such as promoting regional equity and reducing regional imbalances, or reducing rural-urban migration. Due to the geographical specificity of infrastructure, under a given setting, one location could have comparative advantages over another due to the difference in telecommunications infrastructure and could experience a crowding-in effect, and vice versa. This crowding-in effect is very similar to location theory of foreign direct investment that can be used to attract private investment for a particular location.[25]

Empirical evidence on the role of infrastructure as a source of growth in general has been confined to industrialized countries. Empirical evidence on the role of telecommunications in the case of developing countries, such as telephones per worker, shows a positive relationship with growth.[26] In fact, the level at which telecommunications is growth enhancing is the level of at which a country attains universal service obligation.[27]

2.5 Universal Service under Regulated Monopoly and Competitive Market

Similar to most infrastructure services, telecommunications have some public good components[28] and the theorists' justification for their public provision is due to the presumed market failures. Yet it was not until the late 19th century when the worldwide move for such public provision began.[29] However, during recent times, the trend has again reversed and public provision for presumed public good is questioned.[30]

---

[25] See for example Dunning (1980).
[26] See Easterly and Levine (1997) in case of Africa.
[27] See Röller et al (2001) for recent evidence.
[28] See Leff (1984) for a discussion on public good components of telecommunications.
[29] See Lal (1997).
[30] See for example, Hart et al (1997) for a theoretical model that describes why private provision of public good could be superior. The paper contains an application to the prison privatization.

2.5.1 Regulated Monopoly and Universal Service

Local telecommunications services have traditionally been provided by a public or regulated monopoly under a regulated price structure where the goal of USO has been pursued through cross subsidization. Regulators have kept local prices uniform across markets through implicit cross subsidies from other services, such as long-distance calls, and they have provided telecommunications access to high costs areas through generating profit from low cost areas.

However, the economic cost of such cross-subsidization is the loss of allocative efficiency resulting from the market distortions. In addition, there are distributional consequences as well. A regulated monopoly that subsidizes local calls with proceeds from long-distance calls, and high cost geographic areas with proceeds from low cost profitable segments distorts market outcomes and may subsidize high-income users living in high cost areas. In some cases, low-income users might be the providers of such a subsidy, instead of becoming the beneficiaries, as low-income users generating long-distance calls and low-income users living in low-cost areas may subsidize high-income users generating local calls and living in high-cost areas. As a result, such a subsidy may not increase the size of the network; rather, this works primarily as an income transfer program from one group of people to another.

Cross-subsidizing local calls for long distance calls as practiced by regulated monopolies in general is inefficient from other economic points of view as well. The argument is based on price elasticity of demand where a long-distance call has relatively elastic demand compared to a local call, and it is economically inefficient to tax a relatively elastic service to subsidize the price of a relatively inelastic service. In fact, empirical findings show that the elasticity of the local call service is very low, ranging from –0.006 to –0.17, and the price of local services has very little impact on the telephone subscription decision.[31]

2.5.2 Competition and Universal Service

Since the late 1980s, public policy towards network industries in general has focused on liberalization and deregulation. The primary aim of such a policy is to liberalize those parts of an industry where competition is feasible while keeping the residual services under a regulated state monopoly. The transformation of telecommunications from a regulated monopoly to a competitive market brings the question of USO to the forefront. The compatibility of competition and USO is an object of intense political and economic debate in telecommunications. There is a trade- off (trade-off) between competition and universal service provision that is not easy to resolve. Moreover, competition can lead to outcomes that are not socially desirable.

[31] Crandall and Waverman (2000).

One trade-off frequently quoted in economic literature is that competition might lead to 'cream-skimming'. In an economy where the cost of telecommunications provision varies in different regions, laissez-faire will lead to 'too much' entry in profitable regions and 'too little' in the loss-making regions. Under such circumstances, a state or regulated monopoly would not be able to maintain/serve universal service obligations in loss-making regions under uniform pricing requirements due to the unavailability of cross subsidy. Because of these problems, it has been felt that competition and universal service requirements do not mix well.[32]

Empirical evidence based on the US telecommunications liberalization experience usually does not report any trade-off between competition and universal service provision. Studies that are concerned with network expansion usually find a higher growth in subscribership following liberalization of telecommunications.[33] In addition, studies that are concerned with redistribution effects find that liberalization has resulted in higher penetration, and growth rates were higher among lower income households and among regions previously characterized by lower telephone densities.[34]

It should be mentioned that, though a high penetration rate could be an affordability indicator, it is not the sufficient condition of affordability. The basic caveat is that as the price elasticity of demand for telecommunications, particularly for local calls, is very low, households might decide to subscribe to such services even when such a subscription causes a significant budgetary burden on them.

Empirical studies that examine the issue at aggregate level are far from conclusive and there is no definite conclusion as to whether competition is harmful or beneficial to the universal service objective.[35]

2.6 Financing Universal Service

Before the 1980s, universal service had typically been provided by a public or regulated monopoly in industrialized countries and, whether provided or not, the same type of institutional form had been responsible for providing universal service in developing countries as well. As a result, due to the monopolistic nature, the financing mechanism has been designed accordingly. Since the 1980s, as countries have been liberalizing their communication sector, maintaining the financing of USO in industrialized countries, and finding the new financing mechanism in the case of developing countries, has emerged as a

[32] See Armstrong (2001) for a formal model.

[33] See Mueller (1993), the study examines period of liberalization following the expiration of BELL patents in 1893-1894.

[34] See Gillet (1994), the study examines the period following AT&T's divestiture in 1984.

[35] See Barros and Seabra (1999) for a cross-country time-series comparison.

challenge. As state monopolies were often the providers of universal service, such policy reform puts a strain on universal service provision. Under these circumstances, it is necessary to develop alternative mechanisms that provide universal service but do not require explicit public transfer and do not distort the competitive market environment. The two best practices that will be discussed in chapter 4 are the two examples of such mechanisms.

2.7 Universal Service in Practice

Different countries defined universal service in different ways. Due to the focus of the present study, it reviews some of the present practices in developing countries.

Developing countries usually define universal service based on three criteria: 1) population size of a community, for example, a telephone for every settlement of 'x' population; 2) distance of a telephone, for example, a telephone within 'x' kilometres; and 3) travelling time, for example, a telephone within 'x' minutes travelling time. The definition of universal service in developing countries usually includes one or more of these criteria. For example, Ghana defines universal service based on population, 'one telephone in every locality of more than 500 people', in Brazil it is the distance from a telephone 'within less than five kilometres', in South Africa it is 'a thirty minute travelling distance to a phone'.[36]

There are competing approaches with their pros and cons that are practiced to provide universal service in the developing countries context.

2.7.1 Cross-Subsidy

Despite the liberalization, cross subsidy has remained the most prevalent form of universal service funding. In order to maximize subscription/access to telecommunications, universal service mechanisms usually rely on cross-subsidy where subsidies are spread uniformly across households specific to high cost area or to type of users (for instance residential users). The major advantage of cross-subsidy is that the implementation of such a scheme requires minimum regulatory supervision. However, such cross-subsidy is not targeted to those users who, in the absence of subsidy, would drop off the network. It has also been found that such cross-subsidy does not have any significant influence on subscribership.[37] In addition, as the emphasis is on considering USO under a liberalized market environment to attain efficiency, competition and social desirability together, cross-subsidization is not a desirable option in the present context.

---

[36] ITU (1998) p. 69.

[37] See Kaserman et al (1990) in the US context.

2.7.2 Cost of Service Approach
Many countries use cost of service approach either to regulate the incumbent or to provide support for universal service where a carrier is compensated by the additional cost that it needs to incur to provide services to the less profitable areas and to the less profitable subscribers. Similar to many other government actions, the design of a cost based subsidy suffers from imperfect information. If it were possible for a regulator to know the user-specific cost of telecommunications service, then it would have been possible to allocate subsidy based on actual cost. However, such information is either not available or is costly to collect. As a result, this approach suffers from measurement difficulties, disputes over data and methodology, and the interests of the parties, to mention only a few.[38]

2.7.3 Targeted Subsidy
To increase subscribership, targeted subsidy to groups of people who would not subscribe to the network otherwise has been a common practice in industrialized countries. Targeted subsidy has advantages over cross-subsidy as such subsidy is non-distortionary in nature. However, the applicability of such targeted subsidy at household level in the context of developing countries seems unattainable due to the enormous financial burden that such subsidy would put on the public budget.

2.8 Conclusions
Universal service ensures non-discriminatory access and widespread availability at uniform price and acceptable quality. There exist both economic rationales such as positive production and consumption externalities, and development rationales such as the presence of merit goods and income redistribution that call for the provision of universal service. However, due to the very nature of networks and the existence of externalities, competitive equilibria do not exist. This implies that the first-theorem of welfare economics cannot be applied to the telecommunications sector. Due to this, the pursuit of universal service in industrialized countries has been achieved under a state monopoly through cross subsidization. However, such cross-subsidy creates allocative inefficiency and distorts the market outcomes.

The goal of universal service in developing countries is to provide access to telecommunications at a 'reasonable level', termed as universal access. As a result, developing countries usually define universal service based on the population of a community, distance that needs to be travelled to reach the nearest public telephone, and travelling time. There are competing approaches that are practiced in developing countries such as cross-subsidy, cost of service approach and targeted subsidy, which have their own advantages and

[38] See Weller (1999).

disadvantages. However, under these approaches, the goal of universal access for the households that live in rural areas of developing countries seems unattainable.

As the review of the telecommunications sector in Bangladesh and Peru will show, the goal of universal access has not been achieved either in Bangladesh or in Peru. Despite the cross-subsidization of local calls for long distance and international calls, the goal of universal access has never been attained and the state of telecommunications services both in Bangladesh and in Peru has remained underprovided. As a result, it is important to evaluate the best practices that are helping to achieve universal access in the rural areas of Bangladesh and Peru. Consequently, the next chapter reviews the state of the ICT sector and universal access in Bangladesh and Peru, and the subsequent chapter evaluates the two best practices in the two countries.

3 THE STATE OF THE ICT SECTOR AND UNIVERSAL SERVICE IN BANGLADESH AND PERU

3.1 Introduction

Information and communication technologies include two distinct but interrelated technologies: information technology and communication technology. In terms of services, information and communication technologies, henceforth ICTs, include a wide range of services at present and, due to the rapid advancement in technology, this range is expanding rapidly. However, for the review purposes, the current study will concentrate mostly on telecommunications (both fixed and wireless) services. The study will review the state of the Internet in brief due to its limited availability in the present context. The apparent restriction of the review is due to the fact that other forms of services are not widely available in the study context.

The review process proceeds as follows: it reviews the ICT sector in Bangladesh and Peru where the review includes telecommunications sectors and information sectors. The chapter reviews the telecommunications sector in more detail and includes fixed telephony, mobile telephony, public telephony and rural telephony. Due to the very focus of the study, it reviews the state of universal service in telecommunications in rural areas of Bangladesh and Peru. In the case of the information sector, the review includes the state of the Internet in Bangladesh and Peru, and cyber cafes in the case of Bangladesh. The review concludes with drawing some similarities and differences in the ICT sector that exist between the two countries.

3.2 Overview of the ICT sector in Bangladesh

Bangladesh lies in the northeastern part of South Asia between 20°34′ and 26°38′ north latitude and 88 °01′ and 92°41′east longitude (see box-3.1). The country consists mostly of flat fertile land where agriculture is the main occupation of the people, employing 68.5 percent of the labour force and contributing 25.25 percent of the gross domestic product (GDP). As of 1999, the per capita income of the people in Bangladesh was US$362 while 35.6 percent of the population lived below the national poverty line. It is one of the low-income countries according to the World Bank classification. In terms of human development, Bangladesh is classified as a low level of human development and ranked 132 in UNDP's human development index. The country has a parliamentary form of government at present. For administrative purposes, the country is divided into six divisions, which are further subdivided into 64 districts. Box-3.1 provides a political map and some selected indicators of Bangladesh for the year 1999.

Box 3.1: Basic indicators of Bangladesh, 1999

Capital	Dhaka
Population, total (in millions)	127.67
Rural population (% of total)	76.04
Population density (per sq km)	980.79
Mortality rate, infant (per 1,000)	61.20
Life expectancy at birth	58.9
Adult literacy (% of age 15&>)	40.8
GDP per cap., PPP (current int. $)	1483
Population below poverty line (%)	35.6
Agriculture, value added %	25.25
Manufacturing, value added %	14.92
Services, etc., value added %	50.48
Administrative System:	6 administrative divisions
	Parliamentary form of government

Source: (WDI) 2001, UNDP (2001).

Table-3.2 presents some basic indicators related to the information and communication sectors in Bangladesh. The number of telephones (both fixed and cellular) per thousand inhabitants shows the availability of network communication at aggregate level. It is a widely used indicator and is assumed to capture the ability to adopt new innovation such as information technologies. The number of Internet hosts and Internet users are the supply and demand sides of information technologies respectively. These two figures indicate the state of diffusion and adoption of information technologies. The high tech export that stands for high- and medium-technology export as a share of all exports of a country indicates the state of creation of new technologies in general, and information and communication technologies in particular. The last indicator, the mean years of schooling, represents the human skills necessary to adopt and create technologies, particularly information technologies.[39]

Table-3.2: Basic Indicators of ICTs in Bangladesh

Indicators	**State**
Telephone (mainline and cellular per 1,000 people, 1999)	5
Percentage of households with a telephone (approximate)	0.0188
Internet hosts (per 10,000 people)	0.000078
Internet users (per 1000 people, 1999)	0.3916
High tech export [1/]	2.9
Mean years of schooling (age 15 and above, 2000)	2.6

[1/] High- and medium-technology exports as a share of all exports. Source: UNDP (2001), WDI (2001), ITU (2001).

[39] See UNDP (2001) pp. 46-47 for a composite index on ICTs.

However, as can be seen from the table, in terms of the selected indicators, the state of information and communication technologies in Bangladesh has remained at a very low level in general. For instance, the availability of telephones, which is one telephone for each 200 inhabitants and approximately one telephone for each 53 households is inadequate by any standard. This inadequacy is reflected in the case of other indicators as well, as can be seen from the Internet hosts per 10,000 people and Internet users per 1,000 people. In terms of capability to adopt ICTs, particularly information technologies, the country is also lagging behind. With average schooling of less than three years, the country ranks poorly in the human skills necessary for the wide adoption of information technologies. However, despite the low achievements of Bangladesh in terms of all basic indicators presented here, it exports high technology content products that represent around 3 percent of total exports. This last indicator implies that the adoption and diffusion of ICTs in Bangladesh is not evenly distributed and there might be some small technology hubs that produce ICTs. This also shows the inequity in access to ICTs in Bangladesh.

3.2.1 Markets and Regulation

Table-3.2.1 summarizes the market and regulatory structure of information and communication technologies in Bangladesh. As can be seen from the table, the telecommunications sector has been liberalized only partially. The urban fixed telephony market, which is by far the largest market, has still remained under state ownership. In contrast to fixed telephony, the cellular telephony sector has been liberalized where the number of private operators is limited to the number of concessions granted by the government. In the case of the Internet, the sector is under open competition and has by far the largest number of private operators.

Table-3.2.1: Summary of ICT Market and Regulatory Structure in Bangladesh, 2001

Communications service	Market Structure	No. of Licensed operators	Ownership Type
Fixed Telephony-urban	Monopoly	1	State
Fixed Telephony-rural	Partially liberalized[1/]	2	Private
International PSTN	Monopoly	1	State
Cellular Mobile	Oligopoly	4	Private
Radio Paging	Monopoly	1	1
Pay Phones-Urban	Monopoly	1	State
Pay Phones-Rural	Duopoly	2	Private
Internet Access	Open Competition	30	State and Private
Telegraphs	Monopoly	1	State

[1/] The Government has granted one license for the South and one for the North. Source: Various Government Documents, Bangladesh Government.

It appears that there is a negative relationship between the life of an ICT service and the extent of market liberalization; the more concurrent a ICT service is, the more liberalized the market structure it has and the less regulatory constraint it faces. For instance, among the services cited in table-3.2.1, fixed telephony is the oldest service and the Internet is the latest; while fixed telephony is still under either a monopoly market structure, or liberalized only partially, the Internet market is characterized by open competition.

The information and communications sector in Bangladesh is regulated by the Ministry of Posts & Telecommunications (MOPT). The Acts and Ordinances that govern the sector are The Telegraph Act of 1885, The Wireless Telegraphy Act of 1933, The Bangladesh Telegraph and Telephone Board Ordinance of 1979, and The Bangladesh Telegraph and Telephone Board (Amendment) Act 1995. The primary legislation that governs the sector is the Telegraph Act of 1885 that provides the government with the exclusive power to establish and provide all telecommunications services and products.

The Wireless Telegraphy Act of 1933 governs the operation of one-way Radio Communications, Paging and Radio services. The Bangladesh Telegraph and Telephone Board Ordinance of 1979 provides the Bangladesh Telegraph and Telephone Board (BTTB) with the monopoly rights and powers for issuing licenses for telecommunications and wireless services. However, the amendment of 1995 has cut| back some of the rights granted to BTTB under the earlier ordinance, including the exclusive privilege regarding telegraphs and the power to grant licenses. The amendment has also transferred the regulatory authority from BTTB to MOPT. Now MOPT is in the process of forming a regulatory body called the Telecommunication Regulatory Board (TRB). Presently MOPT is the principal telecommunications policy maker. Other institutions, such as the Ministry of Finance and the Planning Commission are also involved in the policy making process.

3.2.2 Review of the Telecommunications Sector

The telecommunications sector in Bangladesh is characterized by low coverage, high cost and poor quality. Though the importance of telecommunications is well recognized in the various documents published by the Bangladesh Government, such documents also express the overall frustration about the real progress.[40] After the independence of Bangladesh in 1971, the communication sector, similar to most other sectors, has followed the course of a state planned development program. The First Five Year Plan was launched in July 1973 and was followed by a Two Year Plan (1978-80) against the backdrop of worldwide

---

[40] See, for example, Bangladesh Economic Review, various issues including 1997, 1998. For instance, Bangladesh Economic Review (1997, p.73), an annual publication of the Bangladesh Government, writes ‘although considerable progress has been made in telecommunications, it is still lagging behind the rate of progress which is fast taking place in other parts of the world or in the region’.

inflation and uncertainties. In 1980, the five year plan framework was reinstated and since then three five year plans in succession have been implemented. There was no development plan during 1995-97 after the expiry of the Fourth Plan (1990-95). Since 1997, the sector and the country have been operating under the Fifth Five Year Plan (1997-2002).

Table-3.2.2 shows the public allocation to the communication sector and actual expenditure under the various planning phases. Contrary to the government's recognition of the significance of the communication sector, the public allocation to the sector has rather decreased in the late 80's and early 90's. One of the major objectives of the on-going five-year plan is to ensure universal telephone services. However, quite contrary to the objective, the allocation to the communication sector has remained 2.5% of total public allocation.

Table-3.2.2: Public Allocation to the Communication Sector and Actual Expenditure (million Taka)

Period	Amount	As a % of total public allocation.	Actual exp. as a % of allocation
1973-78	1180	2.99	90.51
1978-80	1300	3.99	72.31
1980-85	3410	3.07	101.76
1985-90	3880	1.55	164.43
1990-95	6520	1.88	186.04
1997-02(est.)	21865	2.55	

Source: Various Five-Year Plans, GOB.

Compared with Peru, the telecommunications sector in Bangladesh has been liberalized only partially. So far, the state monopoly BTTB has not been privatised. Within the existing market and regulatory structure, private operators are allowed to offer mobile, paging and the radio trunking services in both urban and rural areas. Up to 1989, the state monopoly, BTTB, was the only provider of telecommunication services in the country. In 1989, regulators allowed the private sector participation but only in the case of rural telecommunications. In 1996, the mobile telecom market was liberalized and brought into open competition. However, the urban fixed telephony market has remained under the monopoly rights of BTTB.

Figure-3.2.2 shows the development of the telecommunications sector in Bangladesh, indicating the major changes that have taken place since the mid 80s. Under state ownership, the size of the telecommunications sector as a percentage of GDP had remained 0.3 percent. After the partial liberalization of the rural telecommunications sector in 1989, the share of the sector had jumped to 0.56 percent in 1990. The sector had expanded further to 0.9 percent of GDP in 1997 following the opening of the mobile telephony sector in 1996. However, despite this expansion, the size of the sector has remained less than one percent of GDP. The study reviews the telecommunication sectors under two major categories, fixed telephony and mobile telephony. As the access to telephones has geographic characteristics, it reviews the state of rural telecommunications separately as well. In addition, the current study reviews the public phone and other private initiatives. In reviewing the different services, it concentrates primarily on coverage (supply), price and quality.

Figure-3.2.2: Development of the Telecommunications Sector in Bangladesh

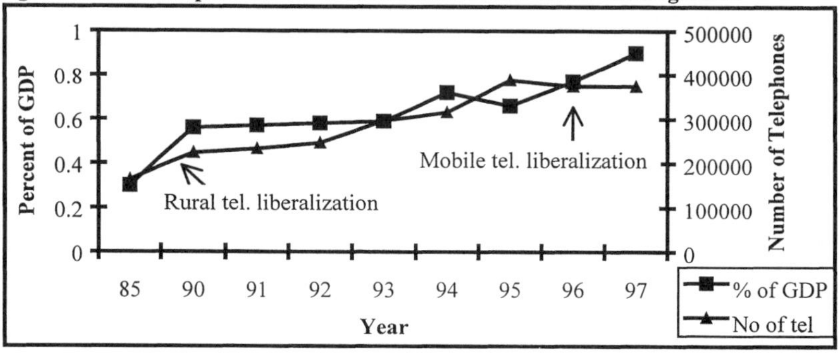

Source: GOB, Statistical Yearbook, various issues.

3.2.2.1 Fixed Telephony

Table-3.2.2.1a provides some basic indicators of the fixed telecommunications sector in Bangladesh. Three years, 1990, 1996 and 1999 have been chosen so that indicators can capture the state of telecommunications under three different regimes: pure state monopoly, after the intervention in the rural telecom market, and after the liberalization of mobile telephony. However, it should be noted that despite these two interventions, the fixed telephony in Bangladesh has basically remained a state owned service.

The level of network development in the fixed telephony in Bangladesh has remained very low. As can be seen from the table-3.3.2.1a, for every 454 inhabitants there was one telephone in 1990. Despite some changes in the telecommunications sector, the situation has improved only marginally and telephone density reached a mere 0.34 percent in 1999. However, the low level of telephone density and penetration in Bangladesh are not due to the lack of effective demand. Despite the high connection charge, which has always remained higher than GDP per capita, the waiting list for main lines has been increasing over time and reached around 42 percent of the existing supply (main lines in operation).

In addition to the limited supply, the fixed telephony under state ownership in Bangladesh has a highly distorted tariff structure and, despite the partial liberalization of the sector, there was no substantial change in tariff structure.

BTTB does not discriminate between business and residential users and charges a very high connection fee. Though it reduced the fee in the mid 90's by around 18 percent, it increased the fee in the late 90's by a further 52 percent. While it charges a very high connection fee, which is one of the highest in the region, the monthly subscription fee is relatively lower compared to other countries.

Table-3.2.2.1a: Basic Indicators of Telecommunications in Bangladesh

Indicators	1990	1996	1999
Telephone density [1/]	0.22	0.26	0.34
Telephone penetration [2/]	1.29	1.47	1.8
Waiting list for main lines	113,656	145,854	181,525 [b)]
Cost of three minute local call (US$)	0.04	0.04	0.03
Residential telephone connection charge (US$)	298	246	375
Business telephone connection charge (US$)	298	246	375
Residential monthly telephone subscription (US$)	4.33	3.58	3.05
Business telephone monthly subscription (US$)	4.33	3.58	3.05
Percentage of digital main lines	11	40	66 [b)]
Telephone faults per 100 main lines	91.08	n.a.	17.32

[1/] Main lines per 100 inhabitants; [2/] Main lines per 100 households (approximate); [a)] 1991 figure; [b)] 2000 figure. Source: ITU (2001), BTTB and GOB.

The distorted tariff structure can be found in the price of fixed telephones as well. The local tariff of BTTB is heavily subsidized: while local calls costs BTTB US$0.032 (Tk.1.59) per minute, it charges US$0.034 (Tk.1.7) per call irrespective of call duration.[41] BTTB compensates the revenue deficit of local calls through cross subsidization, which results in one of the highest tariffs for national long distance and international calls. The minimum tariff for national long distance calls is US$0.068 (Tk.3.4) per minute, and the tariff can be as high as US$0.84 (Tk.42) per minute. For international calls, the tariff structure is from US$0.52 (Tk.26) to US$2 (Tk.100) per minute.[42]

In respect of quality, the fixed telephone network has marked some improvement during the last decade. The percentage of digital lines has reached to 66 percent in 1999 from 11 percent in 1990, and faults per main lines have also decreased to 17.32 percent in 1999 from 91.08 percent in 1990. However, despite these improvements, the quality of service has not improved sufficiently enough. The very high complaints rate, which is more than 50 complaints per 100 lines and the very high percentage of unsuccessful local calls, which is around 80 percent of local calls, show the low service quality that has been persisting in fixed telephony under state ownership in Bangladesh.[43]

---

[41] GrameenPhone Annual Report 1999.

[42] Mentioned tariffs are for 2001, source: BTTB.

[43] See World Bank and BCAS (1998).

It follows that there is an overall backwardness prevailing in the fixed telephony in Bangladesh. However, is it because Bangladesh is a poor country? Or, is it specific to the policy and institutions of the country? To answer these questions, a regional comparison has been carried out. To put the telecommunication sector in a regional perspective the study has selected some South Asian neighbours, presented in table-3.2.2.1b, who are at a similar level of development in terms of GDP per capita. Though there is a high correlation between telephone density of a country with its GDP per capita (both US$ and PPP), compared to its neighbours, Bangladesh should have had a minimum telephone density of 15 lines for every 100 inhabitants in 1999. As can be seen from the table, its neighbours, those who are poorer than Bangladesh, had higher telephone density compared to Bangladesh. This implies that it is not only the low-income that prevents Bangladesh from having a larger network, but that there are some other factors such as institutional arrangements and regulatory structures that are constraining the telecommunications sector development in Bangladesh.

Table-3.2.2.1b: Telephone Density- A Regional Comparison, 1999

Country	Telephone Density [1/]	Cost of a 3 min Local Call (In US$)	Cost of a 3 min Local Call (In PPP)	GDP per capita (In US$)	GDP per capita (In PPP)
Nepal	11	0.01	0.08	222	1237
Bhutan	18	0.02 [a/]	n.a.	512	1341
Bangladesh	***3***	***0.03***	***0.14***	***362***	***1483***
Pakistan	22	0.02	0.08	508	1834
India	27	0.01	0.09	450	2248
Sri Lanka	36	0.05	0.18	814	3279

[1/] Main lines per 100 inhabitants; [a/] 1998 figure; n.a.: Not available; PPP: Purchasing Power Parity; Source: ITU 2001, WDI 2001.

3.2.2.2 Mobile Telephony

Bangladesh entered into the era of mobile telecommunications in 1989 with a private monopoly operator PBTL known as City Cell. The sector turned to a competitive one in 1996 when the regulatory authority allocated nationwide mobile licenses to three more operators. The introduction of competition has worked as a dynamic force and resulted in a rapid network expansion, reduction in access and use tariff, and improvement in service quality. While in 1992 there were only 250 mobile phone subscribers, the numbers reached 283,000 at the end of 2000, representing a more than 100,000 percent growth in less than a decade. The unprecedented growth in mobile subscriptions is matched by an equal drop in connection charges. As can be seen from table-3.2.2.2a, while under the private monopoly operator, the connection charge for mobile phones was US$869, the charge reduced by 96 percent and dropped to US$31 in 1999 under the competitive market. This significant reduction in connection charges is accompanied by a 90 percent reduction in monthly subscription charges and a more than 50 percent reduction in tariff for local calls during the same period. In addition to the increased penetration and reduced cost, the quality of the

network has improved also, as the operators are offering a GSM or CDMA platform as opposed to an AMPS platform.
Table-3.2.2.2b shows the mobile telephony providers in Bangladesh, the technology, market share, and their recent performance. At present mobile operators offer mobile phones, voice mail services (VMS), and short message services (SMS). In addition, one of the operators, GrameenPhone (GP), has added prepaid mobile phone services since September 1999. With 68 percent of the total market share, as of 2000, GP is the current market leader.

Table-3.2.2.2a: Development of Mobile Telephony in Bangladesh under Different Regimes

Indicators	Private Monopoly 1995	Competition 1999
Connection charge (in US$)	868.96	30.55
Monthly subscription (in US$)	49.65	5.09
Cost of a 3-minute local call (peak, in US$)	0.52	0.24
Cellular penetration rate[1/]	0.0021	0.21[a)]

[1)] Cellular subscribers per 100 inhabitants; a) 2000 figure. Source: ITU (2001) and Industry Sources.

Table-3.2.2.2b: Mobile Telephony Market in Bangladesh, 1999-2000

Operator	Technology	2000 A	2000 B	2000 Total	1999 Total	Growth
PBTL	AMPS	5,500	0	5,500	18,400	-70.11
PBTL	CDMA	14,200	13,300	27,500	13,000	111.54
GrmeenPhone	GSM	32,000	160,000	192,000	60,050	219.73
TMIB	GSM	34,000	1,700	36,000	25,274	42.44
Sheba	GSM	16,000	4,000	20,000	11,939	67.52
Total		**102,000**	**179,000**	**283,000**	**130,662**	**116.59**

Note: A: With BTTB's interconnection, B: Without BTTB's interconnection
Source: www.emc-database.com

After the liberalization of the mobile telephony market, the country has two telecom networks at present - one is the fixed telephone network, built, managed and operated by state monopoly BTTB. The other network is the newly established mobile operators' network, which one might consider as a private sector provided network. These two networks are not well connected and interconnectivity between the private mobile operators' network and the public fixed line operator's network (BTTB) is perhaps the biggest hurdle to overcome in the telecommunication sector at present. Despite the impressive growth of mobile telephony, due to the relative scarcity of the BTTB's interconnection, not all mobile subscribers are connected to BTTB's fixed network (table-3.2.2.2b), and a mobile phone with access to BTTB's network costs much more than one without such access.

The new call traffic created by the cellular operators has created acute interconnection problems with the fixed network, as the fixed network is not sufficiently equipped to absorb the additional traffic.[44] Due to the insufficiency of interconnectivity, 179,000 mobile owners have remained isolated from the 616,000 fixed telephone users of BTTB. To overcome this bottleneck, mobile operators of the country have offered an investment plan to expand BTTB' infrastructure. However, the public authority concerned has yet to approve the plan.[45]

The aggregate duty, including value added taxes, on new handsets is about 60 percent of import duties, which is one of the highest in the region. In addition, the annual royalty on mobile phones is US$22 (Tk.1,100). This tax structure can be considered regressive, as the average cost of a mobile phone is half of the cost of a fixed phone. In a country with a very low penetration rate, a very high fixed phone connection charge, and a long waiting list for fixed lines, households who otherwise could not have a phone are obtaining mobile phones and are subject to the payment of these high taxes and duties. Needless to say that unprecedented growth of the mobile telecom sector has offered a handsome windfall for the government's revenue collection.

Despite the explosive growth in mobile telephony and an associated increase in penetration rate, the new network has also remained urban biased. As with fixed line operator BTTB, mobile operators concentrate mostly in two major cities of Bangladesh, Dhaka and Chittagong. With the exception of GrameenPhone, all the other three operators have a customer base only in these two cities. However, GP, the current market leader, has expanded its network to 30 district towns including six divisional headquarters. Additionally, it has a plan to cover the whole country in the near future.

3.2.2.3 Public Telephony and Private Call Centres

In a country of around US$300 average per capita income, the ability to pay US$375 connection charge for a fixed private telephone is limited primarily to the rich part of the population that mostly lives in urban areas. Under such conditions, the publicly accessible telephone has an important role to provide connectivity to the poor people living particularly in rural areas. However, contrary to the expectation, the total number of public telephones in 1991 was 1,812 and the number reached 2064 in 1999. In terms of relative availability, this makes less than one public telephone per 6000 inhabitants!

To improve the public telephone services, Bangladesh Telephone Silpa Sangstha, BTSS (Bangladesh Telephone Industrial Corporation), another state-owned enterprise involved in telecommunications equipment production, introduced card phones in 1992. The card phone in Bangladesh is now

---

[44] GrameenPhone Annual Report 1999.

[45] The Financial Express, January 15, 2001.

synonymous with the public phone as non-card public phones are mostly non-functional. Table-3.2.2.3a shows the growth of card phones over the years and table-3.2.2.3b shows the geographic spread of card phones. After almost a decade since the introduction of card phones, the total number has remained at 1403 in 2000, or one card phone per 100,000 inhabitants. Similar to fixed telephony, the card phone has remained very much urban biased in terms of geographic coverage.

Table-3.2.2.3a: Growth of Card Phones over the Years

Year	Number	Growth
1997	265	
1998	1316	3.97
1999	1355	0.03
2000	1403	0.04

Source: Bangladesh Economic Review 1999, 2000.

Table-3.2.2.3b: Distribution of Card Phones, as of 2000

Location	Number of Card Phones
District	1029
Thana	320
Growth Centre	54
Total	1403

Source: Bangladesh Economic Review 2000, pp. 90.

One recent development in the urban and semi-urban areas in Bangladesh is the establishment of private call centres, which are usually known as phone & fax shops. As the name suggests, such shops resell phone and fax services and some of them, particularly in urban areas, sell email and Internet services as well. Though the number of such centres is not available at country level, based on a primary survey the first approximation suggests that the total number would be around 25,000, of which around 20 percent of them are located at thana level. As thanas are the lowest administrative units situated in rural areas, the spread of such call centres up to thana level is offering a wider access to telecommunication services.

3.2.2.4 State of Rural Telecommunications and VPP Program

Unlike the urban fixed telephony market, the rural fixed telephony market has experienced a partial liberalization since as early as 1989. The authorities had divided the country into two parts, South and North, and granted a license to two private operators for monopolistic operations. Out of 464 thanas,[46] two private providers, Bangladesh Rural Telecom Authority (BRTA) and Sheba Telecom were granted licenses to operate in 200 thanas in the North and 199 thanas in the South, respectively. The licenses were granted for a 25-year

[46] Thana is the lowest administrative headquarter in Bangladesh.

period. The state-owned monopoly, BTTB kept the rest of the thanas under its own operation control.

Table-3.2.2.4: State of Rural Telecommunications: Before and After Liberalization

Indicators	Public Monopoly 1980	Oligopoly 1998
Total number of telephone mainlines	26,683	38,534
Share of rural area as a % of total telephone	10.50	8.90
Telephone lines per 100 rural inhabitants	0.031	0.039

Source: Industry Sources.

Table-3.2.2.4 shows the state of rural telephony under public monopoly and limited competition. In terms of both relative share and absolute performance, rural telecommunications remained underprovided. In 1980 telephone lines per 100 rural inhabitants stood at 0.015. After almost two decades, the figure has changed to 0.039, one telephone per 2546 inhabitants living in rural areas. During the same period, from 1980 to 1998, telephone lines per 100 urban inhabitants have increased from 0.68 to 1.735. Despite the privatisation of the rural telecom market, the share of rural areas as a percentage of telephones has not increased over the period. In addition, after almost a decade of privatisation of the rural telecom market, the coverage has remained limited mostly to urban centres in rural areas. Rural settlements, particularly villages, have been left without telecommunications.

Between the two rural fixed line monopolies, BRTA has installed more lines than Sheba Telecom. As of December 2000, BRTA had around 20,000 subscribers in 100 thanas. The subscribers included 13,000 fixed lines, including 3000 public call offices, 5,500 wireless loop (WLL) lines, and 1,500 RRT lines. The other fixed telephone operator, Sheba Telecom, as reported in Richardson et al (2000), had only 1,500 subscribers across 195 southern rural thanas.

Village Pay Phone (VPP) Project: To address the issue of rural peoples' access to telecommunications, Grameen Bank (GB) and GrameenPhone (GP) have founded a joint project called the village pay phone (VPP) program. It is a partnership between GB and GP. GB, through its wholly owned subsidiary, Grameen Telecom (GTC), and in collaboration with GP, allows female borrowers of the Bank's credit program to become the owner-operators of mobile phones under the VPP program, where GP offers access to the network at a discounted rate, and GB operates and manages the project. As of March 2001, the program has connected 4,166 villages under its network. However, this program will be discussed in chapter 4 under the alternative mechanisms for provision of telecommunications.

3.2.3 Review of the Information Technology Sector

Access to the Internet includes private subscriptions and access through public places. The example of the first type includes access at home through subscriptions to Internet service providers (ISP), and access through Web-TV. The second type of access includes access at public places such as in libraries, and schools. In between these two extremes characterized by ownership rights and user rights, there are a couple of other possibilities that are a continuum of a mixture of ownership rights and user rights. Examples include access at work and at commercial Cyber Cafés.

The availability of off-line email access to limited users in the capital city of Bangladesh started in 1993. However for on-line services users had to wait until 1996 when the state monopoly BTTB commissioned VSAT[47] and two private providers started VSAT based Internet services. Prior to this, dial-up links were used to communicate to an overseas e-mail server using international phone calls and the uploading and downloading of mails took place using UUCPs (Unix-to-Unix copy). After the introduction of VSAT, the ease of access to the Internet increased exponentially owing to point-to-point direct transmission. The opening up of the Internet market in 1996 had resulted in a competitive market. As of December 2000, there were approximately 30 Internet service providers (ISPs) under private ownership out of which 20 of them were in Dhaka and the rest were in three other major cities: six in Chittagong, two in Sylhet, and two in Khulna. Apart from private ISPs, the BTTB has its own Internet wing serving mostly government entities.[48]

The prerequisite of access to the Internet through direct subscription is the availability of a private telephone and a personal computer. In addition, a user needs to incur both fixed cost, such as a one time connection fee that can vary from US$20 (Tk.1000) to as much as US$100 (Tk.5000), and variable cost such as a fixed monthly charge and usage charge and associated telephone bill. Depending on the time of day and working and non-working days, the usage charge varies from US$0.01 (Tk.0.5) to US$0.06 (Tk.3.0) per minute. Due to the competition among ISPs, there is a downward pressure on usage rates and on connection charges as well.

Table-3.2.3: Basic Indicators of Information Technology in Bangladesh

Indicators	1997	1998	1999
Number of internet hosts	300	1,000	3,000
Estimated internet users	1,000	5,000	50,000

[47] VSAT stands for very small aperture terminal. It is a small earth station with 0.6 to 2.4 meters antenna designed to handle voice, data and video communication.

[48] See Management Consulting Group (MCG) (2001).

Number of personal computers	30,000	120,000	130,000
Penetration rate [1/]	0.0008	0.0040	0.0394
PC per household	0.1364	0.5217	0.5417

1/Internet users per 100 inhabitants; Source: ITU (2001) and Industry Sources.

Table-3.2.3 shows the basic indicators of the information sector for the period of 1997 to 1999. Similar to mobile telephony, the information sector has also experienced a rapid growth in users. As of 1997, the penetration rate was a mere 0.0008, or one Internet user for every 125,000 inhabitants in Bangladesh. By 1999, the penetration rate jumped to 0.0394 or one Internet user for every 2564 inhabitants. However, despite the current surge in use, given the low number of personal computers, the potential number of users in Bangladesh is limited to less than 0.1 percent of the population.

Due to the limited personal computer ownership and low telephone penetration rate, access to the Internet in public places can play a significant role in Bangladesh. One recent development in the urban metropolis to this end in Bangladesh is the cyber café, a café that provides access to Internet and email services. As of January 2001, there were eight cyber cafés in Dhaka: five under private ownership, two under non-government ownership, and one under public ownership. The use tariff varies from US$0.04 (Tk.0.7) per minute to US$0.06 (Tk.3) per minute. The lower use tariff is usually preceded by an access cost that varies from US$4 (Tk.200) to US$5 (Tk.250). However, similar to Internet services in Bangladesh, cyber cafés have remained limited mostly to Dhaka city.

3.3 Overview of the ICT sector in Peru

Peru lies in Western South America, bordering the South Pacific Ocean, between Chile and Ecuador and the geographic coordinates are 10°00′ south and 76°00′ west. The country has a wide variety of climatic zones with rainforests, high mountain ranges and dry deserts. As of 1999, the per capita income of the people in Peru was US$2346 while 49 percent of the population lived below the national poverty line. It is one of the lower middle-income countries according to the World Bank classification. The country has a very high literacy rate and UNDP (2001) has classified Peru as a dynamic adopter of information and communication technologies and ranked 48 in the overall technology achievement index (TAI).[49] In terms of the overall human development index, Peru is classified as a medium level country in human development and ranked 73.The country is a constitutional republic and divided into 24 departments and one constitutional province. Box-3.2 provides a political map and some selected indicators of Peru for the year 1999.

---

[49] TAI is composed of technology creation, diffusion of recent innovations, diffusion of old innovations, and human skills. For more information, see UNDP (2001), pp. 46-49.

Box 3.3: Basic indicators of Peru, 1999

Capital	Lima
Population, total (in millions)	25.23
Rural population (% of total)	27.58
Population density (per sq km)	19.71
Mortality rate, infant (per 1,000)	39.20
Life expectancy at birth	68.5
Adult literacy (% of age 15&>)	98.3
GDP per cap., PPP (current int. $)	4622
Population below poverty line (%)	49.0
Agriculture, value added %	7.17
Manufacturing, value added %	23.651
Services, etc., value added %	55.328
Technology Achievement Index	0.271
Administrative system:	24 Departments 1 constitutional province

Source: WDI (2001), UNDP (2001).

Table-3.3 presents some selected indicators related to the information and communication sector in Peru. The indictors are similar to that of Bangladesh and represent the availability of network communication, ability of the country to adopt information technologies, the state of diffusion of information technologies, creation of information and communication technologies and the human skills necessary to adopt ICTs. One additional indicator in the case of Peru is the gross enrolment ratio at tertiary level in mathematics and engineering during the period 1995-97, which reveals the level of human skills development necessary for the creation of information and communication technologies.

As can be seen from the table-3.3, the state of information and communication technologies in Peru shows a moderate level of achievement in terms of the selected indicators. For instance, the availability of telephones has increased to one telephone for less than 10 persons and approximately one telephone for every two households. In respect of the availability of Internet hosts, the achievement is also reasonable. However, the users have not adequately responded with this increased supply as less than two percent of the population is using the Internet at present. In terms of capability to adopt ICTs, particularly information technologies, the country meets the human resource prerequisites; with more than seven years of average schooling, Peru has a high potential for a wide adoption of information technologies. In addition, with a gross enrolment ratio of 7.5 percent at tertiary level in mathematics and engineering, the country can become a creator of information technologies.

Table-3.3: Basic Indicators of ICTs in Peru

Indicators	State
Telephone (mainline and cellular per 1,000 people, 1999)	107
Percentage of households with a telephone	49
Internet hosts (per 10,000 people)	7.0
Internet users (per 1000 people, 1999)	15.85
High tech export [1/]	2.9
Mean years of schooling (age 15 and above, 2000)	7.6
Gross tertiary science enrolment [2/]	7.5

1/ High- and medium-technology exports as a share of all exports; 2/ Gross enrolment ratio at tertiary level in science, mathematics and engineering, 1995-97; Source: UNDP 2001, WDI 2001, ITU 2001.

3.3.1 Markets and Regulation

Table-3.3.1 summarizes the market and regulatory structure of information and communication technologies as of 2001 in Peru.

Table-3.3.1: Summary of ICT Market and Regulatory Structure in Peru, 2001

ICT service	Market Structure	No. of Licensed Operators	Ownership Type
Fixed Telephony-urban	Oligopoly [1/]	6	Private
Fixed Telephony-rural	Oligopoly [1/]	6	Private
International PSTN	Partially liberalized [3/]	2	Private
Cellular Mobile	Oligopoly	3	Private
Radio Paging	Partially liberalized [3/]		Private
Pay Phones-Urban	Oligopoly [1/]	6	Private
Pay Phones-Rural	Oligopoly [2/]	3	Private
Internet Access	Open Competition	54	Private

[1/] Government has granted 6 licenses but only one is in operation at present.
[2/] Three private operators have been allocated six rural areas under a rural telecom development fund.
[3/] Limited to the number of concessions granted by the government.
Source: Industry sources.

As can be seen from the table, in contrast to Bangladesh, Peru has a more liberalized information and communication sector and the extent of state ownership is virtually non-existent. For the fixed telephony, the government has granted six licenses; however, only one operator is in operation at present. In the case of rural telecommunications, the government has established a fund for the areas where private providers do not extend services voluntarily. Under the program, the government has granted three operators licenses to operate pay phones in six rural areas. However, the program will be discussed in more detail in chapter-4. In the case of the Internet, the sector is under open competition and has by far the largest number of private operators.

The telecommunications sector in Peru is regulated by the following major legislations: Decreto Supremo 062-93-TCC sets general telecommunications legislation, Decreto Supremo 062-94-PCM sets an independent regulatory authority, Decreto Supremo 006-94-TCC is the national telecommunications act and Decreto Supremo 020-98-MTC sets the national policy for the liberalization of the telecommunications market. Under the Decreto Supremo 062-94-PCM, the country has established an independent regulator, OSIPTEL, after its Spanish acronym (the *Organismo Supervisor de la Inversión Privada en Telecomunicaciones*), for the telecommunications sector. Following the legislation, OSIPTEL was created in 1994. The legislation gives OSIPTEL technical, economic, financial, functional and administrative autonomy, and it (OSIPTEL) reports directly to the president of the country.[50]

The primary function of OSIPTEL is to promote a competitive environment and to ensure interconnections among carriers. It oversees tariffs and ensures equal treatments between incumbents and new entrants. In addition it operates and manages the rural telecommunication development fund.

3.3.2 Review of the Telecommunications Sector

The telecommunications sector in Peru has passed through some fundamental changes during the decade of the 90s. The principal forces behind these changes are the privatisation of state owned telecommunications providers, market liberalization and changes in consumers' demand. Similar to other countries, Peru had a public monopoly in the telecommunications sector until the mid '90s. Under state monopoly, there were two state owned telecom enterprises, *Compañia Peruana de Teléfonos* (CPT) and *Empresa Nacional de Telecomunicaciones* (ENTEL). CPT was the provider of basic telecom services in the Lima area, while ENTEL was the national and long distance carrier as well the provider of local telecom services in the rest of the Peru.

The structural adjustment program that the Peruvian government introduced in 1990 paved the way for privatisation of telecom enterprises. Divestiture took place in 1994; using a first-price sealed bid mechanism, and both companies, CPT and ENTEL, were sold to the Spanish based venture *Telefónica de España*. Shortly after buying both companies, *Telefónica de España S.A.* merged them and created *Telefónica del Perú S.A.* (TDP). TDP was granted national monopoly for the provision of telephone lines, local calls, national long distance calls (NLD) and international long distance calls (ILD). Although TDP's monopoly right was initially scheduled to expire in June 1999, the TDP reduced the term so as to expire in August 1998.[51]

The monopoly era that ended in 1998 was marked by competition in all segments of the market. In the competition era, two institutions important for

---

[50] See Torero et al (2001).

[51] See Torero (2000).

the telecommunications sector in Peru are the Ministry of Transportation and Communications (MTC) that awards licenses, and the independent regulatory body OSIPTEL that regulates the telecommunications operators and the market. In the competition era, new operators can establish networks to provide local, national and international services. New entrants could also use TDP's existing network by paying an interconnection fee.[52]

Figure-3.3.2 shows the development of the telecommunications sector over the last decade indicating the major institutional changes that took place during that time. Under state ownership, the size of the telecommunications sector in 1992 was 1.2 percent of GDP and the total number of main lines was 613,711. After the privatisation in 1994, the sector had doubled and reached 2.4 percent of GDP and the number of main lines had increased more than 2.5 times and reached 1,555,093.

Figure-3.3.2: Development of the Telecommunications Sector in Peru

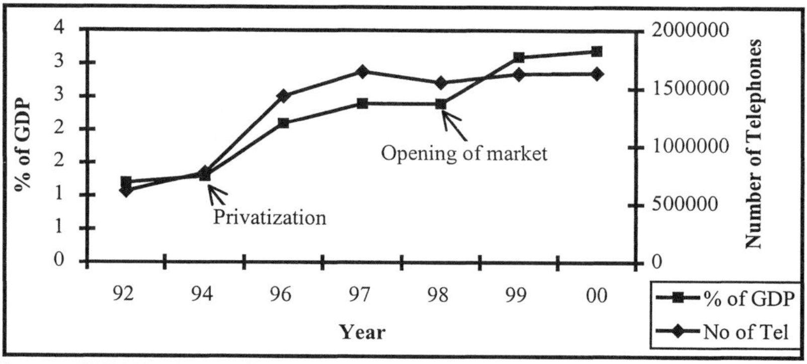

Sources: ITU 2001 data set, industry sources.

3.3.2.1 Fixed Telephony

Table-3.3.2.1a shows some basic indicators of fixed telephony under three different market regimes: state monopoly, private monopoly after the privatisation, and competition. The indicators are chosen so that they can represent the extent of network development, cost of access and use, and quality of service.

In terms of the extent of network development, privatisation of the state monopoly to a private monopoly has resulted in a more than two-fold increase

[52] These proved to be a controversial issue as the interconnection fee charged by TDP was much higher than the neighboring countries. Ultimately, OSIPTEL had to intervene to settle this issue.

in telephone density and telephone penetration rate. Examining the waiting list for main lines, the low teledensity and penetration rate in Peru under the state monopoly were not a result of inadequate demand. The long waiting list, which was more than three hundred thousand just before the privatisation, supports this claim.

Table-3.3.2.1a: Basic Indicators of Telecommunications in Peru under Different Regimes

Indicators	**1993**	**1997**	**1999**
Telephone density [1/]	3	6.75	6.45
Telephone penetration [2/]	17	42	49
Waiting list for main lines	308,247	50,367	29,574
Cost of three minute local call (US$)	0.1	0.09	0.06 [a)]
Residential telephone connection charge (US$)	586	351	131 [a)]
Business telephone connection charge (US$)	1,141	486	131 [a)]
Residential monthly telephone subscription (US$)	3	16	14
Business telephone monthly subscription (US$)	8	20	14
Percentage of digital main lines	38	88	95
Percentage of unsuccessful local calls	59.46 [b)]	1	1 [c)]

1/ Main lines per 100 inhabitants; 2/ Main lines per 100 households; [a)] 2000 figure; [b)] 1992 figure; [c)] 1998 figure; Source: ITU (2001).

Peru had a distorted tariff structure under state ownership of telecommunications. The fixed cost of telephone installation – telephone connection charge – under the state monopoly was one of the highest in the region. In fact while the connection charge was very high, the monthly subscription fee was relatively low. In contrast, the local call tariff was relatively high and so was the international tariff. After the privatisation, the distorted tariff structure has rebalanced significantly and the opening of the market to competition has improved the tariff structure further.

In terms of quality, the state provision represented by the CPT and ENTEL era was below international standards as only around 40 percent of the local calls could be successfully completed. This low completion rate was attributed mainly to the small size of the network and obsolete technology. Under the private monopoly, the percentage of unsuccessful local calls reduced to a mere one percent, and the trend continued during the competitive era.

It is important to note that though the opening of the market and the introduction of competition has not resulted in an increase in penetration rate, all the prices – access cost, fixed monthly tariff, and local call tariff – have marked a significant decrease. In some cases, for instance the telephone connection charge, the cost has decreased by more than 60 percent. In addition, the reduction in tariff has come with an improvement in quality that can be found with the digitalisation of the network.

Table-3.3.2.1b: Teledensity- A Regional Comparison

Country	Telephone Density [1/]		Cost of 3 min local call (US$)		GDP per capita (US$)	
	1993	1999	1993	1999	1993	1999
Bolivia	3.0	6.2		0.09	700	956
Ecuador	5.3	9.1	0.01	0.01	1150	1419
Colombia	11.3	16.0	0.01	0.03	1305	2261
Peru	***2.9***	***6.7***	***0.10***	***0.07***	***1450***	***2346***
Brazil	7.5	14.9	0.03	0.03	2550	4479
Chile	11.0	26.3	0.09	0.09	3035	3613
Mexico	8.8	11.2	0.13	0.13	3880	5121
Argentina	12.3	20.1	0.08	0.09	6910	8100

1/ Main lines per 100 inhabitants, Source: ITU 2001, WDI 2001.

To put the telecommunications sector of Peru in a regional perspective, the present review has selected some countries from the Latin American region based on GDP per capita in 1993. As can be seen from table-3.3.2.1b, among the selected countries Peru had the lowest telephone density in 1993. Though Bolivia, Ecuador and Colombia all had a per capita GDP lower than Peru, they had a higher penetration rate than Peru. Needless to say that countries with a per capita GDP higher than Peru also had a higher penetration rate. It is important to note that the lowest penetration rate in Peru in the region in 1993 under state ownership in telecommunications was accompanied by one of the highest cost (for local calls) in the region, Mexico alone being higher.

After the liberalization in telecommunications, where does Peru stand compared to other countries in the region? To answer this, the state of indicators in 1999 under private monopoly has been examined. In terms of absolute teledensity, Peru has not exceeded any of the countries in the region except Bolivia. However, in terms of relative performance Peru's teledensity has increased by 131 percent between 1993 and 1999, which is one of the highest in the region, next only to Chile. Regarding cost, though Peru has a local tariff higher than the regional average, the local tariff has decreased by 30 percent since privatisation.

3.3.2.2 Mobile Telephony

Mobile telephony in Peru has a liberalized market structure where competition is limited to the number of licenses granted by the government. There are three mobile telephony providers at present: *Telefónica del Perú* (TDP), Bell South Peru and Tim. TDP is the dominant telecommunications provider in Peru and operates by far the largest cellular phone network (in a CDMA platform). Bell South Peru – begun as Peru's first cellular telephone provider Tele2000 a decade ago — the company is now controlled by Atlanta based Bell South. It is now expanding its network (TDMA platform) beyond its Lima base to major provincial cities, which will put it in a competitive position with TDP. In addition, Bell South was awarded a fixed telephony license in 1999 to install a fixed network in Lima and its adjacent port Callao that will also strengthen its competitive position with TDP.

Table-3.3.2.2: Development of Mobile Telephony in Peru

Indicators	1994	1995	1996	1997	1998	1999
Cellular connection charge (US$)	300.68	299.60	n.a.	n.a.	n.a.	n.a.
Cellular monthly subscription (US$)	48.10	47.93	24.54	27.27	29.00	29.00
Cost of 3 minute local call (peak) (US$)	0.81	0.80	1.08	1.04	1.29	1.10
Cellular density /1	0.22	0.31	0.83	1.73	2.99	4.01

1/ Subscribers per 100 inhabitants; n.a: not available; Source: ITU 2001

In terms of performance, the mobile telephony in Peru has experienced a phenomenal growth over recent years. While in December 1990 there were only 2000 subscribers in Peru, by March 2001 the number has increased to 1.4 million subscribers. The geographical spread is also noticeable as the present coverage includes 120 cities under a mobile telephony network. While the monthly fixed charge for a mobile phone has decreased from US$48 in 1994 to US$29 in 1999, the user tariff for local calls has increased from US$0.81 in 1994 to US$1.1 in 1999. Table-3.3.2.2 shows the development of mobile telephony in Peru over the years.

3.3.2.3 Public Telephony

Due to the high fixed cost and widespread poverty, individual subscription for telephones at the household level is limited to the high-income quartile. Under such circumstances, public pay phones could be a viable alternative to connect the majority of the population. However, during the period of state monopoly, the provision of public telephones also remained underprovided when public phones per 100 inhabitants remained at a mere 0.33 percent The privatisation of telecommunications has not only improved the availability of telephones at the household level, but it has also resulted in a wider provision of public phones compared to the previous regime.

Table-3.3.2.3: Public Pay Phones in Peru under Different Regimes

Indicators	State Monopoly, 1993	Private Monopoly, 1997	Competition, 1999
Public pay phones	7,475	38,290	81,253 [a)]
Public phone density 1/	0.33	1.57	3.17

[a)] 2000 figure; 1/ public phone per 100 inhabitants.

Table-3.3.2.3 shows the availability of public pay phones in Peru under different regimes. Under the state monopoly, the total number of public phones stood at 7,475 in 1993. After the privatisation, the public phone grew at a rate of more than 100 percent and reached 38,290 in 1997. The growth of public telephony continued, though at a lower rate, under the competitive regime and reached 81,253 in 1999.

3.3.2.4 State of Rural Telecommunications and FITEL Project

Despite the recent achievements through institutional developments and market liberalization, the telecommunication network has remained confined mostly to the urban areas. Table-3.3.2.4 presents urban rural distribution of telephone lines in Peru. As can be seen, the number of main telephone lines had marked a significant increase between 1992 and 1996, so did the share of the largest city. Despite a four percent decrease in 1996 from 1992, the share of the largest city in main telephone lines remained more than 60 percent. This shows the disparity in telephone access and an urban-biased telephone distribution.

Table-3.3.2.4: Rural-Urban Distribution of Telephone Lines in Peru

Indicators	State Monopoly, 1992	Private Monopoly, 1996	Competition, 1999
Main telephone lines in operation	613,711	1,435,147	1,628,309
Main telephone lines in largest city	415,281	906,817	n.a.
Percent of urban main lines	99	n.a.	n.a.

Note: n.a.= not available; Source: ITU

The regulator of the country has responded to this situation by obliging the private telecom company TDP to provide public phones in rural areas. However, this regulatory intervention was not sufficient as more than 70,000 rural settlements of less than 3,000 inhabitants, which represents around 30 percent of the country's population, was not brought into the public telecommunications network.

In order to bring in the rural settlements that lack a telecommunications network and where the private sector does not extend services under a market incentive, the government of Peru has responded by setting up a telecommunications investment fund, FITEL, after its Spanish acronym (*Fondo de Inversión en Telecomunicaciones*). The telecom regulatory authority, OSIPTEL, supervises and manages the fund. Under the fund, the objective of the authority is to extend telecommunications in selected rural areas where private operators do not voluntarily provide services. However, the functioning and operation of the fund is discussed in chapter 4 in more detail under the heading 'public-private partnership in Peru.'

3.3.3 Review of the Information Technology Sector

Information technology in Peru, namely the Internet, has experienced explosive growth since 1995. The estimated number of Internet users in 1995 (dial-up and dedicated) has been put at around 8,000. By the end of 2000, this figure has jumped to 470,000, a 5775 percent increase in five years. This represents a penetration rate of 1.83 percent as against a mere 0.03 percent in 1995. Table-3.3.3 presents the basic indicators related to the development of the information sector in Peru over time.

Table-3.3.3: Basic Indicators of Information Technology in Peru

Indicators	1995	1996	1997	1998	1999	2000
Number of internet hosts	813	5,192	3,415	4,794	9,230	10,705
Estimated internet users	8,000	60,000	100,000	200,000	400,000	470000
Number of PCs 1/	350,000	450,000	570,322	749,630	900,000	n.a.
Penetration rate 2/	0.03	0.25	0.41	0.81	1.59	1.83
PC per HH	7.40	9.24	11.41	14.42	16.76	n.a

1/ Personal Computer; 2/Internet users per 100 inhabitants; Source: ITU (2001).

In terms of market and regulatory structure, the Internet market has been enjoying open competition in a liberalized regulatory environment. It is interesting to observe that owing to a favourable environment, the number of Internet service providers (ISP) has increased from 2 in 1996 to 58 in 1998. However, since then the number of ISPs has decreased to 54 in 2001 owing to a consolidation of the industry. Despite the recent fall in service providers, the Internet traffic has experienced a rapid growth at a rate of 80 percent during the 1998-2000 period.

However, diffusion of Internet use is limited by the ownership of personal computers and telephone subscription, as traditional forms of Internet access require a personal computer, software and a telephone line. Similar to many developing countries, most of the households in Peru do not own personal computers and many of them are not direct subscribers to a telephone as described in the telecommunications section. Despite the dramatic decrease in computer price and the decrease in the telephone connection charge, the cost all together has remained significant relative to the household income. As of 2000, around a quarter of the households have subscribed to a telephone and less than five percent of them have personal computers. These factors put a constraint on the growth of Internet use and limit the size of the market.

3.4 The State of Universal Service in Telecommunications

This section focuses primarily on the state of universal service in telecommunications in rural areas in Bangladesh and Peru. For this, it has chosen three indicators: first, share of rural area as a percentage of total telephones; second, telephone lines per hundred rural population; and third, telephone lines per hundred urban population. The indicators that have been chosen are expected to capture both the absolute and the relative availability of telecommunications networks in rural areas. As the rural telecommunications market in both countries has experienced all three market regimes: state provision, private monopoly, and competitive market, all the indicators are arranged accordingly.

Table-3.4a: Access to Telecommunications in Rural Areas: Bangladesh

Indicators	Market Regime	State Monopoly	Private Monopoly	Competition[53]
	Year	1989	1995	1998
Share of rural area as a % of total telephone		10.50	8.90	8.90
Telephone lines per '00' rural population		0.031	0.036	0.039
Telephone lines per '00' urban population		1.175	1.640	1.735

Source: Bangladesh Bureau of Statistics, various yearbooks.

Table-3.4a provides the state of indicators of access to telecommunications in rural areas in Bangladesh under three regimes. Similar to other countries, local telecommunications services in Bangladesh had traditionally been provided by a state monopoly under a regulated price structure. However, despite the cross-subsidization of local calls for long distance calls, the goal of the USO had never been attained, and the extent of telecommunications services remained limited mostly to big cities and urban areas. Since the late 80s, public policy of the country towards telecommunications has focused on liberalization and deregulation. As a part of the liberalization process, the authority privatised the rural telecommunications market in the year 1989. However, under the private provision, the access of people living in rural areas has improved only marginally and the relative share of rural areas has declined. Since 1996, the rural telecommunications market has been characterized by a competitive market structure. However, the access has remained at a very low level; as of 1998, the telephone main lines per 1000 rural population stood at 0.39 percent.

Table-3.4b: Access to Telecommunications in Rural Areas: Peru

Indicators	Market Regime	State Monopoly	Private Monopoly	Competition
	Year	1994	1997	1999
Share of rural area as a % of total telephone		0.83	0.87	1.1
Telephone lines per '00' rural population		0.06	0.12	0.16
Telephone lines per '00' urban population		3.7	7.2	7.6

Source: OSIPTEL, LSMS 1994, 1997, 2000

Table-3.4b provides the state of indicators of access to telecommunications in rural areas in Peru under three regimes. The telecommunications sector in Peru has undergone all three different regimes: state monopoly, private monopoly and open competition. Similar to other countries, the country had a public monopoly in the telecommunication sector until the mid '90s. The first stage ended in 1994 when the sector was transformed from a state owned monopoly to a private monopoly through privatisation. The second stage with the

[53] In Bangladesh, the authority has introduced competition in the mobile telephony market in 1996. However, in the fixed telephony, rural telephony has remained under private monopoly structure.

telecommunications sector under private monopoly lasted from 1994 to 1998. After the introduction of competition in 1999, the sector has been experiencing competition in all segments of the telecommunications market. However, despite these changes in regimes, the access of rural people has remained at a low level. Though each shift in regime was accompanied by an increase in penetration rate (telephone lines per '00' population) in rural areas and an increase in relative share of rural areas, both penetration rates reached only to 0.16 percent and relative share to a mere 1.1 percent in 1999.

Based on the three indicators presented above, one can infer that the state of rural telecommunications services both in Bangladesh and Peru has remained underprovided. Though Peru has achieved substantial improvements in telephone penetration rate and reduction in tariff after the liberalization of the sector, such liberalization and introduction of competition was not sufficient to provide telecommunications to the rural areas. The partial liberalization of the market has helped only marginally to improve the penetration rate in urban areas. However, in any case, neither in Peru nor in Bangladesh, has the goal of universal access been achieved.

3. 5 Bangladesh and Peru: Similarities and Differences in ICTs

Table-3.5 summarizes the major similarities and differences between Bangladesh and Peru. Major similarities between the two countries are in the inadequate provision of telecommunications in rural areas, openness in mobile telephony and Internet market structure, and in the recent growth of mobile telephony and use of the Internet. Major differences between the two countries lie in the telephone density, market structure of fixed telephony, Internet penetration rate, and human capital development.

Similarities: Universal access in telecommunications had not been achieved under state provision either in Bangladesh or in Peru. The subsequent privatisation of the state monopoly in Peru and privatisation of the rural telecommunications market in Bangladesh were not sufficient to ensure universal access for rural households in either of the two countries. Later, the introduction of competition in fixed telephony in Peru and mobile telephony in Bangladesh has not solved the problem of universal access either.

In the case of mobile telephony, both Bangladesh and Peru have competitive markets where competition is limited to the number of licenses granted by the regulatory authority. The introduction of competition in the mobile telephony market in Bangladesh and in Peru has brought rapid network growth, extension of the network beyond big cities and unprecedented reduction in access and user tariffs. However, despite the extension of networks for mobile telephony beyond big cities, mobile operators are focused primarily in the urban market segment.

Table-3.5: Bangladesh and Peru: Similarities and Differences

Indicators	Bangladesh	Peru
Income and Poverty		
GDP per capita, PPP (current int.$)	1483	4622
Prevalence of poverty[1/]	35.6	49.0
Adult literacy rate	40.8	98.3
Telecommunications		
Telephone density[2/]	0.34	6.45
Market structure-fixed telephony	Monopoly	Competitive
Market structure-mobile telephony	Competitive	Competitive
Information		
Penetration rate[3/]	0.04	1.59
Market structure	Competitive	Competitive
Rural Telecommunications[4/]		
Telephone lines per 100 rural population	0.039	0.16
Share of rural area as a % of total telephone lines	8.9	1.1

Notes: Figures are for the year 1999. [1/] Percentage of population live below the national poverty line; [2/] Main lines per 100 inhabitants; [3/]Internet users per 100 inhabitants; [4/] For Bangladesh, the figures are for the year 1998. Sources: ITU (2001), Industry Sources, and WDI (2001).

In the case of information technologies, they have a competitive market structure both in Bangladesh and in Peru and these technologies have experienced a rapid growth both in Bangladesh and in Peru owing to favourable environments However, as the traditional mode for Internet usage has sizeable prerequisites such as a personal computer, fixed telephone connection and software that turn access to the Internet into a costly exercise, the growth of information technologies is limited to the upper income quartile of the population both in Bangladesh and Peru.

Differences: In the case of fixed telephony, the major differences between the two countries are in the extent of network development, the quality of the network and in the market structure. In Bangladesh both telephone density and telephone penetration have remained at a much lower level compared to its neighbour and to Peru. The network in Bangladesh has remained relatively backward compared to Peru both in quality expressed in terms of digitalisation and in call completion rate. In terms of market structure, while in Bangladesh it is the state monopoly, in Peru it is the competitive structure. Similar to fixed telephony, Bangladesh has a lower Internet penetration rate compared to Peru.

In addition, due to the lower level of human capital development in Bangladesh compared to Peru, Bangladesh has lower potentials to adopt the available information and communication technologies. In contrast, Peru has a high level of human capital necessary to adopt ICTs, particularly information technologies, and it also has a high level of specific human capital necessary to create new information technologies, which puts the country in a better position compared to Bangladesh.

To conclude, though a competitive market structure in fixed telephony in Peru and a competitive market structure in mobile telephony in Bangladesh is helping towards the growth of telecommunications networks, providing access to rural people, particularly to the poor people, needs to be addressed separately, as a competitive market structure does not voluntarily extend networks to rural areas. To this end, the present study reviews two alternative institutional frameworks that are considered as two best practices in rural telecommunications provision associated with the two study countries, FITEL in Peru and VPP in Bangladesh. The first one, FITEL, provides an alternative institutional framework that shows how government intervention can solve the universal access question for rural areas with minimum distortion. The second one, VPP, provides another alternative institutional framework that shows how universal access can still be achieved when there is no government intervention. The subsequent chapter evaluates them in details.

4 UNIVERSAL SERVICE UNDER ALTERNATIVE PROVISION MECHANISMS

4.1 Introduction

This chapter deals with the question of 'how to provide universal service' in the context of developing countries, and provides two examples of alternative provision mechanisms for telecommunications in the context of rural areas in the two study countries: Bangladesh and Peru. The two alternative mechanisms that this study examines are considered as two best practices in the provision of rural telecommunications in the context of developing countries.[54] The chapter describes and evaluates the two best practices, i) business-NGO partnership in Bangladesh and ii) public-private partnership in Peru that provide access for rural people to telecommunications services. More especially, the chapter accomplishes two tasks: first, it describes the institutional mechanisms – the operational and functioning mechanisms of these two institutions; second, it evaluates these two institutions in terms of attaining universal service goals. As mentioned in chapter 2, the minimum achievement of universal service in telecommunications in rural areas has remained a key public policy issue in developing countries.

Telecom has public good type characteristics,[55] and it is one of the basic components necessary to ensure access to information and communication technologies (ICT); the latter is widely recognized as a strategic factor to reduce poverty and enhance development.[56] As discussed in chapter 2, there are both economic and development rationales of providing universal service to telecommunications as such access ensures both private benefits, such as reduction in information and communication cost, and public benefits, such as creation of positive network externalities. Empirical evidence in the context of both industrialized and developing countries supports the notion that a reliable telecommunications infrastructure can also be growth enhancing.[57]

Despite the presence of strong economic and development rationales, ensuring universal access, particularly to people living in rural areas of developing countries, has remained a major challenge to the policy makers and development thinkers alike. Though the telecommunications sector in most of the countries has experienced different regimes: state monopoly, private monopoly and open competition, these institutions have their own limitations in solving the question of universal service. As described in chapter-3 in the case

[54] See, for example, Brook and Smith (eds.) (2001) that mentions the project in Peru as a best practice.

[55] See Leff (1984) for a detailed discussion on this issue.

[56] For example, see Okinawa Charter (2000). See also Morales-Gomez and Melesse (1998).

[57] See Easterly and Levine (1997) for developing countries and Röller et al (2001) for industrialized countries.

of Bangladesh and Peru, rural telecommunications sectors in both countries had experienced all three traditional institutions; however, the issue of universal service has remained unresolved under these institutions.

The two provision mechanisms that will be discussed is this chapter, Public-Private Partnership in Peru, and Business-NGO Partnership in Bangladesh, are two examples of the innovative use of markets in attaining universal access for rural households, the segment of the population that is usually ignored by traditional market mechanisms. The two provision mechanisms are also examples of two different roles of government: while in Peru government plays an active role to correct the market failure, in Bangladesh government plays almost no role. As a result, these two examples show how government intervention can solve the universal service obligation with minimum distortion and how universal service can still be achieved when there is no government intervention.

The remainder of the chapter proceeds as follows: first, it develops a theoretical framework that describes the limitations of traditional institutions in providing universal service. This is followed by the description of public-partnership in Peru and business-NGO partnership in Bangladesh. Once the partnerships are described, the chapter evaluates the institutions in terms of attaining the universal service goal.

4.2 Theoretical Framework

Economic literature concerned with the question of universal service provision usually recognizes the existence of two geographically distinct markets: an urban market segment, which is profitable, and a rural market segment, which is a high cost area.[58] Under this set up, the challenge is how to ensure universal service to the high cost rural area without distorting the competitive market environment in the urban segment.

4.2.1 Access of Rural Households under Market Provision

A universal service obligation (USO) is typically viewed as the obligation of an operator to provide to all users a 'basic' package of services of 'good' quality and at 'affordable' prices.[59] Traditionally, the pursuit of USO has been achieved under a state monopoly through cross-subsidization. However, the economic cost of cross-subsidy is the loss of allocative efficiency resulting from the market distortions. In addition, there could be some adverse distributional consequences as well.

---

[58] See, for example, Gasmi et al (2000).

[59] See Valletti (2000); see also chapter-2 and the references therein for more details on universal service obligation.

To overcome the distortion stemming from monopoly, the often-suggested policy prescription is the liberalization of the telecom sector. Increase in competitive market forces through privatisation can enhance efficiency, network extension and provide a higher quality of services.[60] However, the transformation of telecommunications from a regulated monopoly to a competitive market brings the question of universal service to the forefront. Market based private providers might not offer the best alternative to a state, or a state-based monopoly for rural areas in particular, as market based solutions might result in under-provision.

The reason for under-provision of telecommunications in rural areas under market conditions is ineffective demand, as revenue per subscriber might not support network extension. However, it does not imply that the extension of telecommunications networks in rural areas does not generate any positive consumer surplus. In fact, it is possible that despite higher willingness to pay per call by a rural subscriber compared to that paid by an urban subscriber, rural areas might remain underprovided.[61] The figure below explains such circumstances more vividly.

Figure-4.1 has two panels. The upper panel combines willingness to pay (WTP) per call and number of calls per subscriber (sub.), while the lower panel combines underlying consumer surplus (CS) per call and total consumer surplus (TCS) per subscriber (sub.). In the lower panel, the TCS curve can be viewed as the individual demand curve, as well as total willingness to pay per subscriber, and the dotted line is the marginal cost of telecom provider which is assumed here as constant per subscriber. The two corners of the horizontal axis of both of the panels 0 and 0′ can be viewed either as income dimension, or as rural-urban dimension, or both. Here 00′ represents total population.

Despite the higher CS per call and associated willingness to pay of a poor/rural subscriber than that of a rich/urban subscriber, as shown in figure-4.1, a private telecom provider cannot extend its network beyond 0′N* as the marginal cost (MC) of network extension is lower than the total consumer surplus and associated total WTP of such a subscriber. That means, out of a total of 00′ population, the competitive market extends the network up to 0′N*, leaving 0N* population without a telecommunications network due to the ineffective demand.

---

[60] Gasmi et al (2000).

[61] For the evidences on rural households' high willingness to pay for access to public telephones, see chapter 7.

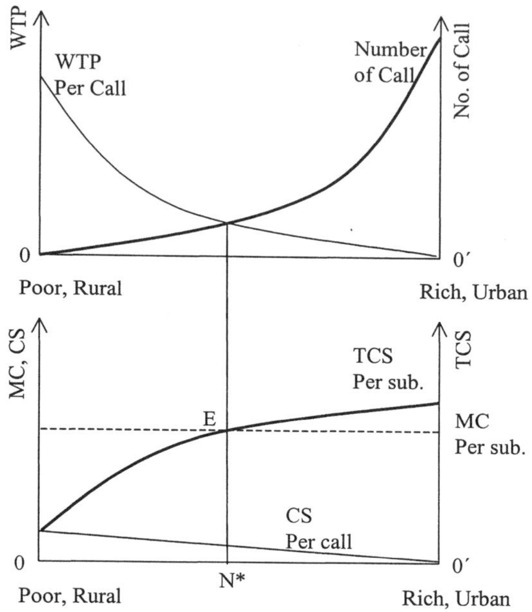

Figure-4.1: Telecom Network under Market Provision

To overcome the ineffective demand, that is the inadequate revenue per subscriber, one possible solution is to sum up the individual demand; demand of the individuals left of N* could be summed up and subdivided in such a way that every individual is included in the network while the marginal cost of inclusion of such a subscriber is less than, or equal to, the marginal revenue from that subscriber. Solutions that are widely used in reality to address part of the 0N* are telephone booths and public call centres. However, as described in chapter-3, network extensions through telephone booths and call centres have also remained underprovided.[62]

4.2.2 Access under Partnership

By partnership the study means to describe a situation where parties involved, e.g., business, government or NGO share cost and/or investment in the production and/or provision of telecommunications services. It is assumed that the production and provision of telecommunications require two inputs, y and z where y is the input from the business and *z* is the input from the government or

[62] For instance, in Bangladesh, as of 1998 the number of inhabitants per public telephone including call centers was 50,000. (Source: author's calculation using ITU' 2000 data set).

from the NGO. If $N(n)$ is the expected output, where n is the number of rural households that have access under partnership, then

$$F(y,z)=\begin{cases} N(n) & if\ z>0 \\ 0 & if\ z=0 \end{cases} \qquad (4.1)$$

Here, $F(y,z)$ is the production function that includes all the services necessary for the provision of rural telecommunications including management and distribution. Note that the input z is essential to provide telecommunication services to the households in rural areas.

Given the above background, it is now possible to formalize the problem. The problem is to maximize the number of persons, n, that have access to telecommunications subject to the restrictions that it does not reduce the number of persons that have access when the market acts alone. Other restrictions are the joint production possibility of the business and of the government/NGO, and revenue and cost considerations. The problem, then, is as follows:

$$\text{Maximize } N(n) \qquad (4.2)$$

Subject to:

$$N(n) \geq N^* \qquad (4.3)$$

$$F(y,z) - N(n) = 0 \qquad (4.4)$$

$$R(n) \geq C(y,z) \qquad (4.5)$$

Here, $N(.)$, is the number of individuals that have access to telecommunications. Equations (4.3) to (4.5) are the constraints: (4.3) restricts the total access to at least at the market provided level, (4.4) is the joint production function where y and z are the inputs of the business and government/NGO respectively, and (4.5) is the revenue and cost restriction which can be viewed as a resource constraint as well. Combining (4.3) and (4.4) into a single constraint, the objective and the constraints can be combined in the Lagrangian expression:

$$L = N(n) - \lambda[F(y,z) - N^*] - \omega[R(n) - C(y,z)] \qquad (4.6)$$

Differentiating with respect to n, y, and z, and assuming no corner solutions, the first order condition (f.o.c) for a maximum is given by:

$$N'(n) - \omega R'(n) = 0 \qquad (4.7)$$

$$-\lambda F_y(y,z) + \omega C_y(y,z) = 0 \qquad (4.8)$$

$$-\lambda F_z(y,z) + \omega C_z(y,z) = 0 \qquad (4.9)$$

The first order condition (4.7) shows the marginal subscriber and the revenue from that subscriber. Conditions (4.8) and (4.9) show the marginal product and marginal cost of the business and government/NGO respectively, where F_y, C_y, F_z, C_z are the respective partial derivatives. The f.o.cs imply that the marginal cost that both the business and the government/NGO incur in order to ensure the inclusion of the marginal subscriber need to be less than, or equal to, the marginal revenue from that subscriber.

4.3 Public-Private Partnership in Peru

Peru offers a unique set of experiences due to its adverse geography and extreme dispersion in the rural population. According to the Population and Housing Census of 1993, around 6 million inhabitants in Peru live in more than 70,000 rural settlements that are characterized by alack of basic infrastructure services such as electricity, water and sewage. So the provision of universal access may lead to the promotion of social and economic development in those areas. In addition, the institutional mechanism that the regulatory authority of Peru has employed to attain universal service in telecommunications for people living in rural areas, the Public-Private partnership, is also unique in the developing countries' context.

In the case of Peru, the relevant concept is universal access, rather than universal service. The Telecommunication Act 1998 that liberalized the market defined the provision of telecommunications based on the principle of equity and extended the right to telecommunications services throughout the country. However, the introduction of competition and subsequent rebalancing of tariffs eliminated the possible source of cross-subsidy for universal access/rural areas.

To ensure universal access, Peru has followed a two-pronged policy. First, contractual obligation with the operators; during the issuing of licenses, the telecom authority of the country included a clause that obliged operators to install and maintain public phones in all localities with more than 500 inhabitants by the end of 1999. For instance, through the concession contract, the Peruvian government obliged *Telefónica del Peru* (TDP) to install 1.1 million additional lines according to a regional distribution, and to install at least one public telephone booth in more than 1,500 villages across the country with more than 500 inhabitants.

Second, the setting up of a development fund named FITEL after the Spanish acronym (Fund for Investments in Telecommunications, FITEL, *Fondo de Inversión en Telecommunicaciones*, or FITEL). The basic objective of the fund is to finance the operation of telecom services in places where private operators do not provide services. As the regulatory obligation on private operators was not sufficient enough to ensure access for all rural people, it was necessary to find additional mechanisms to finance telecom services in rural areas and places of major social interest.

FITEL can be viewed as a Public-Private partnership where public authority and private telecommunications operators work together to achieve a public objective, the so-called universal access. Under this partnership the public authority finances universal access (part of the cost) and private market provides the access. The introduction of competition and subsequent rebalancing of tariffs eliminated the possible source of cross-subsidy for rural areas. To fill the gap, the authority has taken the policy of direct subsidy for rural areas and responded by setting up the fund, FITEL. The primary objective of the fund is to provide access for rural people to public telecommunications services where the private sector acts as a telecommunications provider. The target of the authority is to install at least one public phone for each village. In addition to voice telephony, the public phone has capacity for fax and data transfer at low speeds. However, before describing the program in Peru, the possible role of the private sector in public good provision and auction as a process in public good provision is reviewed in brief.

4.3.1 Private Provision of Universal Service through Auction

Private sector involvements through a competitive environment in public good provision can be welfare improving in general; as such involvements create private incentive for efficiency and innovation in public service delivery.[63] During the last two decades, the 1980s and 90s, reforms that sought to improve the efficiency of public services both in industrialized and in developing countries have resulted in various ways of private sector involvement in public good provision such as change in ownership from public to private through privatisation, contracting out specific services to the private sector, and quasi contracts, to mention only a few. Experiences across a wide range of countries and sectors that have transferred the responsibility of provision of public service to the private sector have shown that such transfers have typically improved efficiency in service delivery and resulted in greater availability of services.[64]

For the regulatory authority in Peru, the issue is how to design policies to ensure universal service and to promote competition where possible so that it is possible to maximise the gains from universal service and yet to retain almost all the benefits of a competitive environment. That means, for a public authority that wants to maximize total social welfare, the issue is how to design policies to enable competition in telecommunications and yet extend universal service so that benefits of both competition and universal service can be attained

[63] See Hart et al (1997) for a formal model and an application for private provision of prisons..
[64] For a host of recent examples, see Brook and Smith ((eds.) 2001).

simultaneously. The problem is to frame universal service in such a way that efficiency, competition and commercial operation are encouraged.

In Peru, the regulatory authority has approached this problem through the private provision of universal service where private participation is ensured through the use of auction in universal service provision. Though the idea of using auction to allocate subsidy for universal service is a relatively new approach, the auction itself is a widely used process in government procurements. It provides the opportunity to reconcile the conflict between the use of a competitive market and the provision of universal service, it minimizes the distortion caused by intervention, facilitates efficient entry and promotes investment.[65]

When it is necessary to compensate for universal service for areas where the private provider does not voluntarily provide the services under price and quality restrictions, auction offers a suitable alternative. The objective of a universal service auction is to minimize the subsidy through competitive bidding among carriers. Auction shifts this information problem from the regulator to carriers and provides a means of revealing the carriers' valuation of the universal service obligation. However, in order to call an auction for universal service, it (universal service) must be clearly defined and auctionable.[66]

Auctions for universal service are usually viewed as a 'Carrier Of Last Resort' (COLR) auctions; i.e., auctions whose object is the obligation to offer pre-specified services at an agreed price. There is a wide range of circumstances in which COLR auctions can reduce the cost of universal service provision compared to more traditional universal service schemes described in chapter 2.[67] However, instead of one for all auctions, repeated auctions over timeare important in order to adjust the universal service package and associated support due to change in technology, cost of universal service production and demand for universal service. In addition, the presence of more than one universal service provider can promote ex-post competition among the carriers.

4.3.2 Operation and Functioning of the Program

The functioning of the projects involves the following steps:

i) Defining the service and selection of a project area:
ii) Selection of an operator
iii) Monitoring the selected operator

---

[65] In a relatively early paper, Demsetz (1968) showed that auction can achieve second best average cost pricing under natural monopoly.

[66] See Weller (1999) for a full treatment of the issue.

[67] See Sorana (2000) for a formal proof.

4.3.2.1 Defining the Service and Selection of a Project Area

Service definition includes the type of services that have to be provided, service standards, and service coverage. At the present stage, services that are provided include at least one public pay phone in each settlement/locality providing access to local and long-distance voice and narrow-band data communications. In addition, the service package also includes one point of public access to the Internet at district level. Once a project is signed, the operator is obliged to provide the agreed services for a 20-year period.

Project areas are usually selected by OSIPTEL based on expressed local demand and the net present value of a project. However, the selection process also takes local authorities' opinions into consideration and includes criteria such as population coverage, existence of education and health facilities, centrality of a location, and relative isolation in terms of telephone access. However, once a project is selected, it needs to get the Ministry's approval before its implementation.

4.3.2.2 Selection of an Operator, Tariffs, and Incentives

Once the Ministry has approved a project, the next step is to select a private operator. For this, the regulator, OSIPTEL, issues an international request for proposals and arranges an auction for minimum subsidy. The auction process is a single round sealed bid. The private bidders are required to provide ex ante financial guarantees that ensure the seriousness of their offer and to prevent 'winner's curse'. A private bidder can choose one or several rural areas or combine them. Under this process, given that an operator satisfies minimum technical specifications, service quality and reliability requirements, an award is made to the operator who requests the lowest subsidy through an open bid. The subsidy covers all items involved in the service provision.

The tariff is regulated by OSIPTEL under a price cap regime that maintains parity with the urban areas. The support amount that an operator receives is region specific. Financial contracts specify the subsidy amount between OSIPTEL and a selected operator resulting from the auction process. The contract usually specifies the terms and conditions of fund disbursement, tying it in with project implementation and service quality: 35 percent of the fund is paid at the start of the project, 25 percent of the fund after the instalment of the facilities, and the remaining 40 percent in half-yearly instalments over a five year period, subject to compliance with the terms and conditions.[68] When a selected operator fails to comply with the service mentioned in the contract, the regulator adjusts the support amount based on the clause included in the contract.

[68] See Cannock (2001), p. 17.

Once an operator is selected, it builds the network, operates, and maintains it. The life of a concession is 20 years, and the operator gets the ownership of the network. Apart from delivering local telecommunication services specified in a contract, the operator is also obliged to provide interconnection with other rural villages on long distance services. OSIPTEL regulates the private operators and ensures the continuity of the service delivery.

4.3.2.3 Monitoring the Operators and Future Adjustments

The regulator needs to monitor the quality and availability of the service at a specified level. For this, there is a built-in mechanism. OSIPTEL supervises project implementation and monitors the operation through a network management system, and requires a dedicated data circuit in the operators' headquarters to oversee service quality and billing. Based on the observations, the OSIPTEL staff prepare a half-yearly report that works as the basis for half-yearly payments of the support funds (from FITEL) to the private operators.

In the case of an operator not complying with the terms specified in the contract, there is a pre-specified penalty that the regulator can impose on the operator. For instance, the regulator reduces the half-yearly instalments by US$1000 per day for pay phone and network monitoring outages, and by 10 percent per locality per week of delay in initiating service for up to one month, at which point the balance of the subsidy is penalized.[69]

The contract with the private operator also specifies other indicators of performance such as: grade of service (network congestion in peak hours), time to get a dial tone, and overall service quality measured by mean opinion scores.

4.3.3 Projects under the Partnership

Up to March 2001, three rounds of competitive bidding have been taking place where four or more bidders have been participating. Accordingly, the authority has assigned six projects of rural telephony under FITEL that will translate into 8,653 rural telephones and will benefit 4,237 rural settlements covering approximately 4 million rural inhabitants. Under the program, the objective of the authority is to cover 5000 rural settlements by 2003.

The distribution of the projects and the probable direct and indirect beneficiaries are shown in table-4.3.3. Out of the six, five projects are in the process of implementation. Gilat to Home is in charge of the North Project (Amazonas, Cajamarca and Piura) and the Centre West Project (Huanuco, Junin, Urna, Pasco and Ucayali). Cand G-Avantec has the concession for the Centre North (Ancash, La Libertad and Lambayeque) and finally Telerep has the concession for the South (Arequipa, Moquegua, Puno and Tacna), Centre-

[69] See Cannock (2001), p. 17.

South (Apurimac, Ayacucho, Cusco, Huancavelica, Ica and Madre de Dios) and the North Jungle (Loreto and San Martin) projects. According to information provided by FITEL, however, Gilat to Home has recently acquired the projects from Telerep.

Table-4.3.3: Rural Communication Projects under Public-Private Partnership in Peru

Project	Settlements	Direct Beneficiary Population	Indirect Beneficiary Population
North1/	938	519957	499114
South2/	534	135917	249468
South Centre2/	1029	303260	528734
North Jungle2/	374	141621	187424
North Centre3/	582	317648	363682
West Centre1/	780	259668	343930
Total	4237	1678071	2172352

1/ Gilat to Home; 2/ bid to Telerep; 3/ CyG Telecomm Avantec
Source: OSIPTEL_FITEL

There are two pilot projects currently in operation (Box-4.3, dots are the settlements with public telephone). The first project, that of the North Borderline Project, comprises the instalment, operation and maintenance of 1 public telephone in each of 193 rural settlements. The fund awarded for the instalment cost is US$ 3.727.379, for operating and maintaining for 20 years is US$ 998.465, and the project fee that the project will have to pay in the fifth year is US$4.909.292. It may be mentioned that the requested amount for operation and maintenance of the net is paid out in half-yearly quotas over 5 years. In this way the winning offer in actual value is US$ 1.661.563, which is 41 percent less than the originally budgeted amount of US$ 4.053.000, which represents a public investment per inhabitant of US$ 11. To have an idea of the impact and benefit of the project, it can be mentioned that the execution of this project will considerably decrease the distance to the nearest telephone for the communities, and will increase the proportion of the population with telephone access in that zone from 48 percent to 88 percent.[70]

The second project is the one for the South Borderline. This project comprises the installation, operation and maintenance of 1 public telephone in 252 rural villages of Tacna Department and the border districts of Puno and Madre de Dios Departments. In terms of impact, the execution of this project will effect a considerable reduction in the distance to the nearest telephone and will increase the proportion of the population with access to the telephone in that zone from 68 percent to 85 percent.[71]

[70] OSIPTEL-FITEL (1999).
[71] OSIPTEL-FITEL (1999).

Box-4.3: Projects in Operation in Peru: North and South Borderline Projects

North Borderline Project | South Borderline Project

4.3.4 Sustainability of the Program

The project depends neither on additional public finance nor on cross-subsidy; contributions collected from the sector are used within the sector. The contributions to the fund come from the telecommunications operators that contribute 1 percent of their annual gross revenue (less tax and interconnection charges) to the fund. Since the inception of the fund in 1994, it has been growing by US$12 million annually.[72] The fund is used exclusively for the financing of telecommunications services in rural areas and localities deemed to be of social importance.

FITEL leverages private investment. As seen in the case of the North Borderline pilot project financed by FITEL, for each dollar of subsidy, the fund has attracted two dollars of private investment. In addition, the administrative cost of the fund is also low, which is less than 2 percent of the collected fund. As FITEL's committed subsidy for the selected projects spreads over a period of five years, the subsidy helps to compensate for the cash flow of the projects at the initial periods. However, once the traffic starts up and grows, the projects are expected to become self-sustainable.

4.4 Business-NGO Partnership in Bangladesh

The public-private partnership described above is an example of attaining universal access for rural areas when the public authority intervenes and corrects the possible market failure. But what if such an authority does not intervene? Is it still possible to use the market in attaining universal access for rural areas? The business-NGO partnership that has undertaken a project named village pay phone program, henceforth VPP program, in Bangladesh provides such an example. However, the description and evaluation of the VPP program

[72] ITU (1998), p. 79.

is preceded by a brief discussion on NGO and public good provision and how an NGO can contribute in a rural telecom project.

4.4.1 NGO and Public Good Provision

The past decade has witnessed a surge in non-governmental organizations (NGOs) in many industrialized as well as in developing countries. One way to see the emergence of this 'third sector' is as a response to "fill" the niches where both the state and the markets have failed. Though many kinds of NGOs exist, the current study is interested in the type defined in Anheier and Salamon (1998): it is an organized sector, institutionally separate from government, and does not return any profits to its owner(s).

Though the economists tend to attribute the production of economic services either to the private sector or to the public sector, the economic contribution of NGOs as a sector is quite significant.[73] In addition to the measured economic contribution, the sector offers more potential, such as innovations in public service delivery and improving the quality of services, and can play supportive roles to the public services.[74] NGOs are considered, particularly by donor agencies, effective to reach and target part of the population and an increasing proportion of both the domestic development budget and foreign aid in developing countries is being channelled through NGOs.[75]

The underlying economic theories that have evolved to explain the rapid emergence of this third sector include demand-side theories, supply-side theories, partnership theory, and social origins approach.[76] And two benchmark models most often used are the public good model and the private consumption model.[77] In the public good model, a desire to increase the provision of public good motivates an NGO to contribute, and thus an NGO would contribute only if it increases the supply of public good to its target group. The current study rationalizes the NGO's contribution in the rural telecom project based on the public good motive. More precisely, in the present framework, an NGO differs from a business (and public) organization in two ways: first, in its concern for increasing the supply of public good; and second, in its information advantage.

---

[73] As found in the study of Salamon et al (1998) on the nonprofit sector in 22 countries of varying income levels, the sector as a whole contributed to 4.7% of GDP in 1995 on an average.

[74] See Badelt (1985); see also Rose-Ackerman (1996) that mentions three important functions that can be played by nonprofit organizations.

[75] For instance, in 1997, the World Bank had most of its environment projects (in 68 countries) implemented through NGOs. See World Bank (1997), pp. 4-5.

[76] For an overview see Salamon and Anheier (1998)

[77] See Roy and Ziemek (2000).

4.4.2 NGOs and Rural Telecom Provision

Provision of universal service in rural areas involves production of services, e.g., building networks, and the operation and management of the service, e.g., subscriber selection. Assuming that these two steps, production, and management and operation of a rural telecom project that intends to provide access for rural people to telecommunications are totally separable, an NGO has a comparative advantage over business in the operation and management of the project in rural areas. This is because the NGO might incur lower operating costs compared to business in the rural context.

Why does an NGO incur lower costs in rural areas compared to a business? Some possible reasons for why an NGO might incur less management/operating cost compared to a business in rural areas are the NGOs presence in rural areas and, related to that, its information advantages. If an NGO has its existing operation and network in a rural area, it does not need to incur substantial set-up costs, which are fixed in nature. By the same logic, the NGO might also need to incur lower variable costs, like operating costs, than a business. In addition, due to its local knowledge, it can overcome the imperfect information problems that exist in rural areas. The empirical evidence that NGOs are more efficient than business are inconclusive. A major problem with comparing the relative efficiency between NGOs and business is that NGOs do not necessarily supply the same good as business.[78]

The argument that an NGO with rural operations may incur lower operating/distribution costs compared to a business is not the sufficient condition for a business-NGO partnership. In fact, the significant difference between an NGO and a business rests on the difference in their objective functions; while businesses are profit driven, NGOs are ideology driven, for instance, increasing the supply of public good for the intended beneficiaries in rural areas.[79]

4.4.3 Village Pay Phone Program Under The Business-NGO Partnership

As described in the theoretical framework, a competitive market that relies on marginal cost pricing cannot extend the telecommunications network beyond N*, leaving ON* rural population without having access to telecommunications (figure-4.1). Now in order to extend access beyond N*, the partnership could set up telephone booths where individual demand can be summed up, or it could establish call centres. However, due to the higher operating and maintenance cost, setting up telephone booths was not a viable option. Therefore it has opted for the call centre and established it under private

[78] See Rose-Ackerman (1996), pp. 721-23, for evidence and a discussion on relative efficiency of business and NGOs.

[79] For evidence that nonprofit organizations are usually ideologically motivated, see James (1982).

ownership. In addition, due to the NGO's public good motive, it wants to ensure access of the rural poor in general and its beneficiaries in particular, and it selects only poor individuals as owners/operators of the call centres as a result.

Grameen Bank (GB) is a non-governmental organization that provides formal credit to rural poor women without requiring any direct collateral, known as group lending.[80] As of February 2000, the bank had 2,355,985 borrowers of which 94.77 percent were female. At its current state, the bank covers around 58.61 percent of all the villages in Bangladesh. As a policy, GB's staff travel everyday to the doorsteps of its borrowers and collect weekly instalments.[81] GB's focus on information and communication technologies could be found in the establishment of three sister organizations: Grameen Communications, Grameen Cybernet and Grameen Telecom.

Grameen Phone (GP) is a commercial venture in Bangladesh. It is a joint venture company that has four shareholders; with a 35 percent stake Grameen Telecom (GTC) is one of its shareholders. Other shareholders include Telenor Invest AS of Norway, Marubeni Corporation of Japan, and Gonophone of the USA. GP was granted a nationwide license for cellular mobile phones in November 1996. GP uses Global System for Mobile (GSM) technology, which is one of the widely accepted digital systems as people in 130 countries use it at present.[82] As a part of its operation, GP establishes networks, and operates and provides telephone connections.

To address the issue of rural peoples' access to telecommunications, Grameen Bank and Grameen Phone have started a joint project named village pay phone (VPP) program. The project, the VPP program, is a partnership between Grameen Bank and Grameen Phone. Under the partnership, the Grameen Phone offers access to the network at a discounted rate, and Grameen Bank operates and manages the project. Grameen Bank, through its wholly owned subsidiary, Grameen Telecom, and in collaboration with Grameen Phone, allows female borrowers of the Grameen Bank's credit program to become the owner-operators of mobile phones under the VPP program.

Under the VPP program, GTC provides cellular phones to the selected borrowers of GB. The members purchase the phone under the lease program of the Bank and make the telephone available to the other villagers. Here each VPP owner-operator acts like a public call office (PCO); she re-sells phone calls, provides message services within the village, and lets others receive incoming calls. Table-4.4.3a shows the growth of VPP over the years. Started in 1998, the VPP program has brought, as of March 2001, more than four

---

[80] See Stiglitz (1990) for the economics of group lending.

[81] See http://www.grameen-info.org/bank/supdates.html

[82] GP's annual report 1999.

thousand rural settlements under its telecommunications network. By 2004, GTC plans to expand the program to all 68,000 villages of Bangladesh. In addition, GTC has a plan to add other products like fax and email services to the VPP program at a later stage.

Table-4.4.3a: Growth of Village Pay Phones over the years

Year	Number of VPP	Growth (in %)
1997	28	
1998	179	539.29
1999	1114	522.35
2000	3085	176.93
2001 (March)	4166	140.16

Source: Grameen Telecom

Box-4.4 shows the coverage of Grameen Phone as of 2001. The VPP program is expanding following the expansion of the network of Grameen Phone. The railway fibre optic that GP uses as the backbone of its physical network covers a distance of around 12 kilometres from any point of the network. Villages after 12 kilometres require additional antennae, and hence additional investment. For such villages, GTC supplies the antenna, GB extends necessary credit, and VPP owner-operators pay additional instalments. However, GP does not commit any additional investment. Table-4.4.3b shows the different components of cost of the VPP packages including the cost of antennae that the VPP owner-operators need to pay to Grameen Telecomm.

Box-4.4: Coverage of VPP in 2001

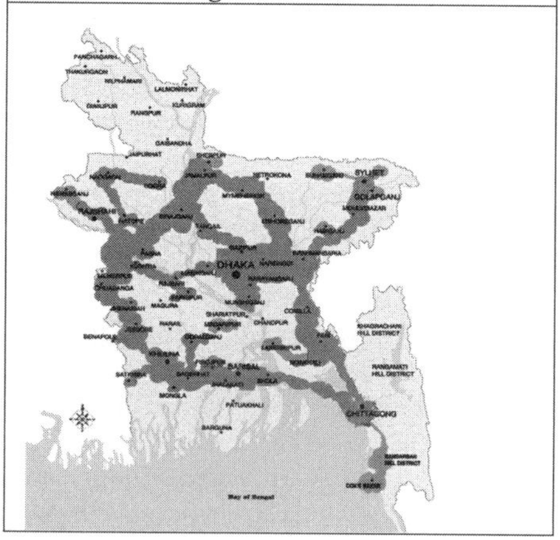

Source: Grameen Phone

Table-4.4.3b: Cost of Village Pay Phone (VPP) Package in Local Currency (Tk.)

Time period	Basic Package	Antenna*	Mobile Set
1997-98	18100	6500	Nokia 1610
1999-Feb'2000	15000	4500	Nokia 1610
March-April 2000	17000	5000	Nokia 8110
May-Dec. 2000	15000	5000	Nokia 5110

*If necessary. Source: Grameen Telecom. Exchange Rate in 2000 (US$1=Tk.50).

4.4.4 Operation and Functioning of the VPP Program

The screening and selection of VPP owner-operators: GB staff select VPP owner-operators exclusively from among credit group members. In the selection process, GB uses financial criteria, such as the borrower's track record of credit and repayments regularity, and financial solvency, such as type of business and flow of income. The technical requirements that GB seeks are literacy, particularly familiarity with Roman numerals; availability of electricity connection in the house; and the centrality of the operator's location within the village. In addition, GB also checks whether a potential operator is willing to offer the service around the clock and go to users' doorsteps to provide message services within the village.

Based on these criteria, GB officials prepare a preliminary list of group members who want to become VPP operators, and all members of the group, in the presence of local GB officials, jointly select one VPP operator for the respective village. GB officials offer the phone to the selected VPP operator through the bank's lease program. Similar to group lending, the group as a whole remains liable for the credit, and members of the group monitor the performance of the VPP operator. At the beginning, GB offers the phone only for three months and monitors the performance and, if satisfied, it then continues further. As part of the program, GTC trains the operators as well as handling all service related issues, and GB collects the bill and monitors the VPP operators through peer monitoring.[83]

Pricing policy: GTC buys bulk airtime from GP at a discounted rate and passes it to the VPP operators. VPP owner-operators pay at the same rate that GTC pays to GP. In addition, operators also pay 13 percent on the airtime charge, to cover GTC's overhead costs and GB's collection costs. VPP owner-operators also charge different rates for incoming calls depending on the type of call. It is important to note that the users who make calls from VPP pay at the same rate that other private GP subscribers pay. That means, under given circumstances, a person willing to make a call is indifferent between calling from a VPP or making a call from his/her private mobile phone (when she/he is a direct subscriber of GP's mobile phone).

[83] See Besley et al (1993) for more on peer monitoring.

4.4.5 Cost and Information Aspect

4.4.5.1 How does the Partnership Keep Cost Low?

In the theoretical framework described above, constant marginal cost per subscriber has been assumed; as the market extends telecommunications to rural areas, the marginal cost per additional subscriber remains the same as for the urban subscriber (figure-4.1). However, empirical evidence shows that marginal cost per subscriber increases as the telecom provider moves from urban to rural areas and this requires either subsidy for rural areas as described in the case of Peru or a higher access price for the people living in rural areas compared to the access price that people pay in the urban areas. However, as will be shown later in this chapter, the VPP program ensures price equality between both urban and rural areas, and between both subscribers and non-subscribers without any subsidy from the government or a cross-subsidy from the urban area. How does the partnership keep the cost low?

There are two types of costs related to the VPP program: the cost of building the physical network in rural areas and the cost of operating and managing the program. For the first, while the building of the physical network is the responsibility of the business – GP under the partnership program – the cost related to operating and managing the program is borne by the NGO – GB. However, as described under the operating and functioning of the program, Grameen Phone does not incur any additional physical investment for the project. For the second, to keep the marginal operating cost at a constant level due to additional subscribers, the partnership uses GB's existing network that keeps the operating and collection cost at a minimum. As the NGO (GB) already has extensive rural operations, the operation and management of the VPP program does not require substantial cost. In addition, to expand the telecom network in rural areas at an affordable price, it is also necessary to create effective demand. To create new subscribers, the partnership has transformed the characteristics of the product; it has transformed the phone from a consumption to a production good and offers it to a class of subscribers, VPP owner-operators, who otherwise would not demand it. To achieve this, the partnership offers credit through the GB's lease program to this new class of subscribers so that they can finance their new investment.

4.4.5.2 How does the Partnership Overcome Imperfect Information?

The approach that the business-NGO partnership has adopted in telecommunications provision is unique in the selection of subscribers, which are the VPP operators. The selection process is unique and also important as both economic success – creating effective demand while keeping the cost low, and public good objective – providing access for rural people to telecommunications, and ensuring ownership of telecom to the rural poor, depend on the performance of the VPP operators. It should be mentioned that when a business alone wants to ensure access for rural poor people to telecommunications utilizing the same mechanism, it faces all three classical problems of imperfect information: screening, incentives and enforcement problems in selecting private owners/operators for its centres. Due to the difference in the likelihood of performance of different individuals, it is costly for a business to evaluate the risk associated with each individual and design a different contract accordingly. This high cost argument is true for the other two as well; it is costly for a relatively new entry to design incentives for the individuals to maximize the payment likelihood, and enforcement of contract in a rural setting for such an entity could prove to be difficult and costly also.

How does the partnership overcome problems related to imperfect information? For this, the partnership relies on the NGO; as the NGO has local level knowledge and interlinkages with other markets, with the credit market in the present case, it does not face the problems stemming from imperfect information. As the program selects from GB's borrowers only, it does not face problems related to imperfect information, as borrowers are well known to GB. In addition, as GB has credit linkage with VPP operators, monitoring the performance and implementing the contract is relatively cheap.

The VPP operators are part of the multilateral debt contract of Grameen Bank's micro credit program where the selection of the VPP operators, the monitoring of their performance, and enforcement of contracts are done by other group members. Such multilateral contracts can solve part of the problem stemming from imperfect information. Specifically, such multilateral contracts can address the screening problem, as group members are known to each other due to frequent social interactions, and each member of a credit group may have better information about other member's efforts and abilities than does the business.[84]

The repayment incentive comes from two sources: first, a default member/group does not have access to micro credit. Given the fact that such borrowers do not have access to formal credit markets, and interest rates prevail in the informal credit market, for example, the interest rate that the money lender charges is relatively high, the NGO's sanction could be a significant penalty for such a member/group. Second, a default member/group may lose the social connections through defaults. Such a penalty might come in the form of denial of access to community facilities, informal credit among others. The multilateral credit contract under group lending provides an opportunity to use non-market institutions to design repayment incentives of individuals and groups.[85]

[84] See Stiglitz (1990) for the likelihood of repayment in a group-lending scheme.

[85] See Besley and Coate (1995) for the use of non-market institutions in group lending.

4.4.6 Sustainability of the Program

To examine the sustainability, the study estimates the profit (loss) from each of the providers that it earns from the VPP program under partnership. The providers of the VPP program include credit group members of the Grameen Bank, those who are the owner-operators of the phone, the network and access provider, Grameen Phone, and the manager of the program, Grameen Bank and Grameen Telecom.

VPP owner-operators: For this group, profit estimation is based on two primary surveys conducted jointly by Jahangirnagar University (JU), Bangladesh, and the Centre for Development Research (ZEF), Germany. The first survey was conducted in June-August 1998; it included 50 VPP owner-operators and 356 users. The second survey was conducted in June-August 2000 and followed up part of the 1998 survey sample; it included 100 users and 24 VPP owner-operators. The details of the two surveys can be found in Bayes et al (1999) and Bayes et al (forthcoming). Table-4.4.6 shows the distribution of weekly revenue and profit of VPP operators in 1998-2000.

Table-4.4.6: Distribution of Weekly Revenue and Profit of VPP Operators (in Tk.)

	1998		2000	
Revenue	1076.82	(417.92)	1575.80	(2233.28)
Profit	591.29	(213.92)	1314.80	(2115.68)

The numbers in the parentheses are the standard deviations (Exchange rate in 2000: US$1=Tk.50.00).

In calculating profit, the labour cost, interest payment and depreciation have been taken into account in addition to the airtime and service charge payment. Depreciation is estimated at 10 percent per year on the total payment that a VPP operator needs to make, and the interest rate is assumed to be 23 percent per year. With a mean profit of Tk.591.29 per week, that can vary from as low as Tk. 230 to as high as Tk. 1149, the VPP ensures a positive return to its operators. Note that in a country like Bangladesh, where per capita income is US$270, the reported mean profit of US$615 is an attractive return to a family that lives close to the poverty line. In addition, as can be seen from the table, the revenue and the profit for the year 2000, the village pay phone has remained a permanent income source for the owner-operators over the period.

GrameenPhone (GP): GP does not incur any additional investment for the VPP program; it offers the access for the VPP owner-operators at a discounted rate. Despite this lower rate, the VPP program ensures a positive return to GP. Two specific reasons that explain this apparent contradiction are: high airtime usage of VPP, and minimum operating cost. As mentioned in GP's annual report 1999, the average airtime usage of VPP is more than double that of a typical business user in urban areas. As GB takes care of the whole operating and collection issues, GP treats the whole VPP program as a single user and incurs a minimum operating cost only. Therefore, as written in GP's annual report 2000, even with discounted prices, the VPP program has proven to be a profitable business for it.

GB and GTC: For GB, the new program neither requires any significant physical investment, nor increases its operating cost, as GB does not expand its office network or human resources due to the VPP program. As of December 2000, GB has earned a net profit of Tk. 867,000.00 from the VPP program. Though the exact figure for GTC is not known, the VPP program is a profitable venture for it as well.

4.5 How Far are the Partnerships in Providing Universal Service?

How effective are these alternative institutions described above, Public-Private partnership in Peru and Business-NGO partnership in Bangladesh, in universal service provision? To evaluate the change in access and change in use due to these alternative provision mechanisms, the following two questions are asked:

i) Does the public-private partnership/business-NGO partnership comply with universal service restrictions?
ii) Does availability of a publicly accessible telephone at village level increase use?

For the first question, the analysis is based on three broad criteria of universal service: access, quality, and uniform pricing as described in chapter 2. The major changes that have been taking place in access due to these two institutions can be analysed by analysing these restrictions.

The second question is relevant in the present context, as mere availability at village level does not ensure that households are using telephones. As opposed to a private telephone, access to a public telephone does not have a direct access cost to individual users, such as a connection cost and a monthly subscription charge as in the case of a private telephone. However, access to a public telephone has non-tariff costs such as travelling time to the public telephone and travelling costs. Given non-tariff costs, does availability at village level make a significant difference in use? Answering this question is important in order to evaluate the relevance of village level availability of public telephones.

Before proceeding further, where does the evaluation of a universal service provided either by a public-private partnership under auction or a business-NGO partnership under a joint project take place? Should it take place at the regulator level or at the users level? However, the problem at regulator level is that it does not have control over total price (tariff plus the subsidy) and hence on social welfare as the price depends on the market outcome that the regulator cannot fully control. The regulator may be unable to observe the quality of service as well, as it might be costly to verify the state and it is the users not the regulators who are the ultimate receivers of the service. These imply that the evaluation of any universal service may be based at the users level.

4.5.1 Universal Service Restrictions and the Partnerships

To evaluate whether the alternative provision mechanisms comply with access, quality, and uniform pricing restrictions of universal service, the study relies on the user's opinion that has been collected from primary surveys in Bangladesh and Peru. It compares all the three restrictions between settlements with and without public telephones so that one can understand the difference that the public telephone provision is making in universal access. For each of the criteria, the study has chosen some specific indicators described below.

For access, the chosen indicators are: distance to the nearest public telephone in kilometres, time required to reach the nearest public phone, transport means to reach the phone, alternative means used to communicate if there were no public telephone, and average expenditure in alternative means of communication as indicators. For the quality of service, the users'_opinion of the present service, and for the price, the non-tariff cost of access such as transport cost to reach the nearest phone, and the value of travelling time have been chosen as indicators.

4.5.1.1 Public-Private Partnership and Universal Service

Access to Telecommunications Services: As one can expect, a priori, the availability of a telephone at village level reduces the distance and time required to reach the public phone. The mean distances for a village with a telephone and a village without a telephone are 2.1 kilometres and 21.5 kilometres, respectively. The mean time required to reach the nearest telephone from a village with a telephone and a village without a telephone is 0.018 hours and 0.587 hours, respectively. The differences in means of distance of telephone and time between a village with a public telephone and a village without a public telephone are statistically significant (table-4.5.1.1a).

Table-4.5.1.1a: Mean Distance and Time to the Nearest Public Telephone: Peru

	Total		Village with phone		Village without phone		Mean Difference[1/]	
Distance of nearest telephone (in km)	12.9	(34.0)	5.19	(38.22)	13.56	(38.16)	-8.362**	(2.426)
Time to the nearest telephone (in hrs)	0.31	(0.89)	0.02	(0.15)	0.59	(1.184)	-0.569**	(0.054)

[1/] Mean difference between village with phone and village without phone; Values in the parentheses are the standard deviations. ** Significant at 1% level.

Table-A.4.5.1.1a to table-A.4.5.1.1f in the Appendix-II show the findings related to the access to telecommunications services in rural Peru in more detail. With the exception of around three percent of the surveyed households, the overwhelming majority use community centre and public telephones as the access point for telecommunications (table-A.4.5.1.1a). The importance of public phones as the access point is reflected in the case of both types of villages, and in the case of both the poor and not poor.

Quality of Service: For the quality of service, the survey asked the users three questions: first, how do they rank the phone service offered by the public phone established under the FITEL program; second, what is their opinion of the service; third, what are the disadvantages of the current phone service. Table-4.5.1.1b shows the users' opinion of the present phone service (Other tables are given in Appendix-II). As can be seen from the table, about half of the surveyed households consider the present service provided under the FITEL project is better than average (good) and the overwhelming majority consider that the service is not poor.

Table-4.5.1.1b: How will You Rank the Telephone Service?

		With Phone			Without Phone		
	TOTAL	Total	Poor	Not poor	Total	Poor	Not Poor
Excellent	13.1	13.6	14.4	12.9	12.4	11.8	13.2
Good	49.6	45.7	49.8	42.1	56.6	59.4	52.3
Regular	33.7	37.6	33.5	41.2	26.6	26.6	26.5
Poor	2.7	2.9	2.4	3.4	2.4	1.7	3.3
Very Poor	0.9	0.2	0.0	0.4	2.1	0.4	4.6

Price: The maximum tariff that an operator can charge under the FITEL project is set by the regulator, and the regulation ensures that the operators comply with the tariff cap. So the tariffs in rural areas are similar to that of urban areas, and uniform pricing between urban and rural users is enforced through regulation. However, there are non-tariff costs such as transport costs and value of travel time reported in table-4.5.1.1c. It is important to note from the table that the non-tariff costs for the households that live in villages with public telephones are not zero. As expected, the non-tariff costs for the households from villages without telephone are much higher than their counter group and the difference in means between the two groups is significant.

Table-4.5.1.1c: Mean Transport Cost and Value of Travel Time to the Telephone (in LCU): Peru

	Total		Village with phone		Village without phone		Mean Difference[1/]	
Transport cost	0.98	(3.50)	0.07	(0.68)	1.87	(4.72)	-1.80**	(0.22)
Value of Travel time	0.23	(0.68)	0.02	(0.15)	0.43	(0.91)	-0.41**	(0.04)

[1/] Mean difference between village with phone and village without phone; Values in the parentheses are the standard deviations. ** Significant at 1% level.

4.5.1.2 Business-NGO Partnership and Universal Service

Access to Telecommunications Services: Table-4.5.1.2a shows the mean distance and time to the nearest telephone. The other indicators of access are presented in the Appendix-II. The mean distances for a village with a telephone and a village without a telephone are 1.14 kilometres and 8.59 kilometres, respectively. The mean time required to reach the nearest telephone from a village with a telephone and a village without a telephone are 0.23 hours and 0.99 hours, respectively. The differences in means of distance of telephone and time between a village with a public telephone and a village without a public telephone are statistically significant (table-4.5.1.2a). It is important to note that the introduction of the VPP program has changed the access of rural households in general; though every village does not have a telephone yet, the access of households from villages without a telephone has also changed due to the introduction of a telephone in a nearby village.

Table-4.5.1.2a: Mean Distance and Time to the nearest Public Telephone: Bangladesh

	Total		Village with phone		Village without phone		Mean Difference[1/]	
Distance to nearest telephone (in km)	4.89	(5.39)	1.14	(0.87)	8.59	(5.42)	-7.45**	(0.46)
Time to the nearest telephone (in hrs)	0.61	(0.59)	0.23	(0.21)	0.99	(0.59)	-0.76**	(0.05)

[1/] Mean difference between village with phone and village without phone; Values in the parentheses are the standard deviations. ** Significant at 1% level.

Quality of Service: Table-4.5.1.2b shows whether the users of VPP are satisfied with the current level of service. In contrast to Peru, users in Bangladesh are less satisfied with the quality of service. One major problem reported in the survey is the poor call completion rate. However, as the quality of the fixed network operated by the state owned BTTB has remained a major problem, improvement in call completion cannot be attained by the VPP program alone.

Table-4.5.1.2b: Are You Satisfied with the Current Level of Service?

	Poor	Not Poor	Total
Satisfied	42.86	45.16	44.79
Dissatisfied	42.86	19.35	23.18
Extremely Dissatisfied	14.29	35.48	32.03

Pricing: Though the regulator of the country does not require it, the VPP program ensures tariff equality between both urban and rural areas, and between both subscribers and non-subscribers. However, as the program operates in a cellular network, the tariff is higher than the fixed network that subsidies the local tariff for long distance and international tariff. Turning to the non-tariff costs, the difference between mean transport cost and value of travel time between the households that live in villages with a telephone and the households that live in villages without a telephone is significantly different (table-4.5.1.2c).

Table-4.5.1.2c: Mean Transport Cost and Value of Travel Time (in LCU): Bangladesh

	Total		Village with phone		Village without phone		Mean Difference[1/]	
Transport cost	10.28	(9.97)	4.54	(6.03)	15.94	(9.85)	-11.41**	(0.97)
Value of Travel time	41.76	(35.72)	17.02	(16.51)	66.15	(32.67)	-49.13**	(3.08)

[1/] Mean difference between village with phone and village without phone; Values in the parentheses are the standard deviations. ** Significant at 1% level.

4.5.2 Public Telephone and Rural Households' Telephone Use

Does the availability of a public telephone at village level have any impact on a household's telephone use? Table-4.5.2 shows the availability of telephones at village level and the households' telephone use in Bangladesh and Peru. While the mean telephone use of the surveyed households that live in villages without a telephone in Bangladesh and Peru is 0.303 and 0.819, respectively, mean telephone use of households that live in villages with a telephone is 0.362 and 0.882. As can be seen from the table, the difference in means between villages with and without telephones is statistically significant both in Bangladesh and in Peru (italic).

Table-4.5.2: Availability of Public Telephone at Village and Household's Telephone Use

Country/ Telephone Use Status	Settlement Status Without telephone		With telephone		Total/ Mean Difference	
Peru						
Do not use telephone	122	(24.5%)	58	(11.8%)	180	(18.22%)
Use telephone	376	(75.5%)	432	(88.2%)	808	(81.78%)
Mean use[1/]	*0.755*	*(0.431)*	*0.882*	*(0.323)*	*-0.13***	*(0.024)*
Bangladesh						
Do not use telephone	108	(75.5%)	90	63.8%	198	(69.72%)
Use telephone	35	(24.5%)	51	36.2%	86	(30.28%)
Mean use[1/]	*0.245*	*(0.431)*	*0.362*	*(0.482)*	*-0.117**	*(0.054)*

1/ The numbers in the parentheses are the standard deviations; ** Significant at 1% level; significant at 5% level.

To see whether the observed difference between the usages of households of two types of villages still persists once controlled for other factors, the study attempts an econometric estimation. For econometric estimation, the approach is to augment a standard cross-sectional telephone use function to include a dummy indicating whether a settlement has a public telephone or not. Let T_i represent a dummy variable that equals 1 if the ith settlement has a public telephone, and zero otherwise. Household j's telephone use decision is assumed to depend on T_i, and a vector of observed characteristics X_{ij}, and can be described by a latent variable model:

$$Y_{ij}^{*} = \alpha' T_i + \beta' X_{ij} + \varepsilon_{ij} \qquad (4.10)$$

where Y_{ij}^{*} is the expected utility level that a household attains from telephone use, ε_{ij} is a normally distributed error terms with mean zero and unit variance, and α and β are parameters to be estimated.

However, one cannot observe the expected utility level; what one rather observes is whether or not a household uses a telephone. Defining for each household j a dummy variable Y_{ij} that expresses the bivariate state of a household's telephone use status, the conditional probability that for household j given T_i and X_{ij}, one observes $Y_{ij} = 1$, (use telephone) is given by

$$P(Y_{ij} = 1 | T_i, X_{ij}) = P(\alpha' T_i + \beta' X_{ij} + \varepsilon_{ij} \geq 0) = F(\frac{\alpha' T_i + \beta' X_{ij}}{\sigma}) \qquad (4.11)$$

The study estimates the probabilities of telephone use given T_i and X_{ij}, where T_i and X_{ij} include a settlement's public telephone availability and household characteristics respectively. It assumes that the accumulated density function is normal; therefore it estimates the equation (4.11) as a probit model.

The equation (4.11) has been estimated with varying sets of covariates both for the VPP program in Bangladesh and for FITEL in Peru. The first specification includes a settlement's public telephone status, household expenditure, distance to the nearest public telephone, time required to reach the nearest public telephone, and distance-time interaction. The estimation has included distance and time to the telephone to account for non-tariff access cost. Note that it has not included a tariff as the tariff does not vary across settlements, and households do not face different sets of tariff. The second specification includes all the covariates of the first specification, together with the distance to the nearest post office as an additional covariate. The rationale behind the inclusion of post distance is to account for the cost of alternatives to a public telephone. The third specification includes all the covariates of the second specification and household characteristics in addition. Table-A.4.5.2a and table-4.5.2b in the Appendix-II provide the summary statistics of the covariates for Peru and Bangladesh, respectively, included in the estimation.

4.5.2.1 Rural Households' Use of Public Telephones in Peru

Table 4.5.2.1a provides probit coefficients of rural inhabitants' use of the public telephone service in rural Peru and table-4.5.2.1b provides the associated marginal effects. As can be seen from the tables, the estimated coefficients remained consistent over all three specifications, and all the signs conform to the prior expectations. Turning to the research question, the availability of a public telephone increases a household's telephone use by 13.52 percent (table-4.5.2.1b, column 2). The impact remains at a level of 13.55 percent once alternatives to public telephones, in addition to income and non-tariff costs, are included, and the impact of public telephones at village level remains at 11.69 percent once household characteristics are included (column 3).

Table-4.5.2.1a: Rural Inhabitants' Use of Public Telephone Service in Peru, Probit Model

Dependent Variable: Use telephone or not (Standard errors are in the parenthesis)

Explanatory Variables	Model 1		Model 2		Model 3	
Public telephone (village with	0.6847		0.6857		0.6129	
telephone=1, no telephone=0)	(0.1051)	**	(0.1052)	**	(0.1099)	**
Household expenditure	0.00012		0.00012		0.0001	
	(0.00002)	**	(0.00002)	**	(0.00002)	**
Household expenditure square	-1.71e-09		-1.71e-09		-1.37e-09	
	(5.27e-10)	**	(5.27e-10)	**	(5.45e-10)	*
Distance to the nearest public	0.0109		0.0108		0.0114	
telephone (in KM)	(0.0038)	**	(0.0038)	**	(0.0040)	**
Travel time to the nearest public	1.1612		1.1422		1.1141	
telephone (in hours)	(0.2366)	**	(0.2369)	**	(0.2353)	**
Distance-Time interaction	-0.0092		-0.009		-0.009	
	(0.0030)	**	(0.0030)	**	(0.0030)	**
Distance to the nearest post office			0.004		0.0025	
			(0.0085)		(0.009)	
Household characteristics:						
Household size					0.0382	
					(0.0303)	
Gender of the head of the					-0.5637	
household (male=1)					(0.1872)	**
Age of the head of the household					-0.0349	
					(0.0259)	
Age square of the head of the					0.0005	
household					(0.0003)	~
Occupation of the head of the HH					0.0853	
(agriculture=1, others=0)					(0.1197)	
Education of the head of the					0.068	
household					(0.0154)	**
Constant	-0.3075		-0.3119		0.2165	
	(0.1275)	*	(0.1280)	*	(0.6041)	
Log likelihood	-385.1713		-385.0428		-369.7634	
Number of observation	988		988		979	
Wald chi2	167.65		167.91		194.83	
Probability>chi2	0.0000		0.0000		0.0000	

** Significant at 1% level; Significant at 5% level; ~ Significant at 10% level.

Table-4.5.2.1b: Rural Inhabitants' Use of Public Telephone Service in Peru, Probit Marginal Coefficients, *Dependent Variable: Use telephone or not (Standard errors are in the parenthesis)*

Explanatory Variables	Model 1		Model 2		Model 3	
Public telephone (village with	0.1352		0.1355		0.1169	
telephone=1, no telephone=0)	(0.0219)	**	(0.0219)	**	(0.0220)	**
Household expenditure	0.00025		0.00025		0.00002	
	(4.15e-06)	**	(4.15e-06)	**	(4.34e-06)	**
Household expenditure square	-3.36e-10		-3.36e-10		-2.60e-10	
	(1.06e-10)	**	(1.06e-10)	**	(1.06e-10)	*
Distance to the nearest public	0.0021		0.0021		0.0022	
telephone (in KM)	(0.0007)	**	(0.0007)	**	(0.0007)	**
Travel time to the nearest public	0.2282		0.2247		0.2117	
telephone (in hours)	(0.0400)	**	(0.0403)	**	(0.0393)	**
Distance-Time interaction	-0.0018		-0.0018		-0.0017	
	(0.0006)	**	(0.0006)	**	(0.0005)	**
Distance to the nearest post office			0.0008		0.0005	
			(0.0017)		(0.0017)	
Household characteristics:						
Household size					0.0073	
					(0.0058)	
Gender of the head of the					-0.0811	
household (male=1)					(0.0658)	**
Age of the head of the household					-0.0066	
					(0.0049)	
Age square of the head of the					0.0001	
household					(0.0001)	~
Occupation of the head of the					0.0165	
HH (agriculture=1, others=0)					(0.0201)	
Education of the head of the					0.0129	
household					(0.0030)	**
Observed P	0.8178		0.8178		0.8161	
Predicted P	0.8830		0.8828		0.8884	
Log likelihood	-385.1713		-385.0428		-369.7634	
Number of observations	988		988		979	
Wald chi2	167.65		167.91		194.83	
Probability>chi2	0.0000		0.0000		0.0000	

** Significant at 1% level; Significant at 5% level; ~ Significant at 10% level.

4.5.2.2 Rural Households' Use of Public Telephone in Bangladesh

Table 4.5.2.2a provides probit coefficients of rural inhabitants' use of the public telephone service in rural areas of Bangladesh, and table-4.5.2. 2b provides the associated marginal effects. The presence of VPP in a settlement increases the telephone use by 36.92 percent (table-4.5.2.2b, column 2). The impact remains at a level of 32.50 percent when the estimation controls for alternatives to VPP in addition to income and non-tariff costs, and the impact of VPP at village level remains at 26.18 percent once the estimation controls for household characteristics are included (column 3).

Table-4.5.2.2a: Rural Inhabitants' Use of Public Telephone Service in Bangladesh, Probit Model

Dependent Variable: Use telephone or not (Standard errors are in the parenthesis)

Explanatory Variables	Model 1		Model 2		Model 3	
Public telephone (village with	0.9928		0.8762		0.727	
telephone=1, no telephone=0)	(0.3171)	**	(0.3345)	**	(0.3615)	*
Household expenditure	8.81e-06		8.67e-06		5.60e-06	
	(0.0000)	**	(0.0000)	**	(0.0000)	~
Household expenditure square	-4.55e-12		-4.44e-12		9.25e-13	
	(02.58e-12)	~	(2.60e-12)	~	(9.62e-12)	
Distance to the nearest public	0.0352		0.0424		0.04	
telephone (in KM)	(0.0478)		(0.0486)		(0.0507)	
Travel time to the nearest public	-0.0093		-0.0116		-0.0157	~
telephone (in hours)	(0.0073)		(0.0076)		(0.0081)	
Distance-Time interaction	-0.0002		0.00001		(0.0002)	
	(0.0006)		(0.0006)		(0.0006)	
Distance to the nearest post office			-0.0715		0.0861	
			(0.0676)		(0.0705)	
Household characteristics:						
Household size					0.0983	
					(0.0605)	
Gender of the head of the					-0.4492	
household (male=1)					(0.4148)	
Age of the head of the household					-0.0115	
					(0.0452)	
Age square of the head of the					0.00003	
household					(0.0004)	
Occupation of the head of the					-0.5108	
HH (agriculture=1, others=0)					(0.2000)	*
Education of the head of the					1.1013	
household					(0.3636)	**
Constant	-1.097		-0.924		-0.9752	
	(0.2641)	**	(0.3087)	**	(1.2758)	
Log likelihood	-137.9588		-137.3865		-124.459	
Number of observations	284		284		284	
Wald chi2	72.40		73.54		99.40	
Probability>chi2	0.0000		0.0000		0.0000	

** Significant at 1% level; Significant at 5% level; ~ Significant at 10% level.

Table-4.5.2.2b: Rural Inhabitants' Use of Public Telephone Service in Bangladesh, Probit Marginal Coefficients

Dependent Variable: Use telephone or not (Standard errors are in the parenthesis)

Explanatory Variables	Model 1		Model 2		Model 3	
Public telephone (village with	0.3692		0.325		0.2618	
telephone=1, no telephone=0)	(0.1186)	**	(0.1276)	**	(0.1385)	*
Household expenditure	2.95e-06		2.89e-06		1.80e-06	
	(0.0000)	**	(0.0000)	**	(1.06e-06)	~
Household expenditure square	-1.52e-12		-1.48e-12		2.97e-13	
	(8.58e-13)	~	(8.61e-13)	~	(3.10e-12)	
Distance to the nearest public	0.0118		0.0142		0.0128	
telephone (in KM)	(0.016)		(0.0162)		(0.0163)	
Travel time to the nearest public	-0.0031		-0.0039		-0.005	~
telephone (in hours)	(0.0024)		(0.0025)		(0.0026)	
Distance-Time interaction	-0.0001		2.522e-06		0.00005	
	(0.0002)		(0.0002)		(0.0002)	
Distance to the nearest post office			-0.0239		-0.0276	
			(0.0225)		(0.0227)	
Household characteristics:						
Household size					0.0316	
					(0.0196)	
Gender of the head of the HH					-0.1602	
(male=1)					(0.1203)	
Age of the head of the HH					-0.0037	
					(0.0145)	
Age square of the head of the					0.00001	
HH					(0.0001)	
Occupation of the head of the					-0.1649	
HH (agriculture=1, others=0)					(0.0654)	*
Education of the head of the					0.2586	
household					(0.0569)	**
Observed P	0.3028		0.3028		0.3028	
Predicted P	0.2765		0.2755		0.2549	
Log likelihood	-137.9588		-137.3865		-124.459	
Number of observations	284		284		284	
Wald chi2	72.40		73.54		99.40	
Probability>chi2	0.0000		0.0000		0.0000	

** Significant at 1% level; Significant at 5% level; ~ Significant at 10% level.

As shown in the case of both Bangladesh and Peru, the availability of a public telephone in a settlement increases a household's telephone use, and the difference in a household's telephone use between a settlement with a public telephone and without a public telephone is significant. Though the inclusion of additional covariates reduces the size of the impact of public telephones, the public telephone dummy variable continues have a sizeable impact in the case of both countries. It may be mentioned that, under each specification, the size of impact of public telephones (VPP) in Bangladesh is higher than in Peru. One possible explanation is that compared to Peru, Bangladesh has a much lower telephone penetration rate and perhaps fewer alternatives to public telephones.

It should be noted that the model that has been used here has not taken into account any additional benefits that households might derive from the use of telecommunications. Such benefits are considered in chapter 6 (and subsequent chapters) in estimating the direct welfare gain that such households derive from the use of the newly available telephone over traditional alternatives.

4.6 Conclusions

The chapter has shown that if there are two geographically distinct markets characterized by two different types of users in terms of revenue generation, users of the segment whose per capita revenue generation is less than marginal cost of telecommunication network extension may not be covered under an unregulated competitive market structure. This outcome may take place even if willingness to pay per call is higher than the cost per call. The primary reasons are the low demand of rural poor households and the high fixed cost of network extension in rural areas and, in some cases, adverse geography.

However, to correct this socially undesirable outcome, the direct intervention of governments through production of services may not be the best alternative. As the experiences of the study countries have shown, such direct state provision may not be sufficient to extend the network to rural areas. Instead, government can ensure market provision through auctions for universal access for areas where the market does not voluntarily extend services. In the absence of government intervention, universal access can still be achieved through market provision given that some entity, for instance an NGO, provides the necessary inputs required for rural areas.

It should be noted that the success of the business-NGO partnership described in this chapter might require the pre-existence of a successful NGO, such as Grameen Bank. However, the existence of any organization with local knowledge is the necessary condition and the public good objective of such an organization is the sufficient condition to provide universal access in rural areas in cooperation with a business/public entity that usually serves the urban segment. However, this generalisation is based on the assumption that there will be no contractual problem between such two entities.

Two projects described and evaluated in the chapter are the examples of innovative use of markets in universal service provision through additional input either from government or from an NGO. Empirical evaluation of the two projects has shown that the projects comply with the criteria of universal service obligation, such as quality, availability (non-discriminatory access) and uniform pricing, both in the case of Bangladesh and in Peru. In addition, the availability of public telephones at village level has also increased a household's telephone use.

One important difference is that FITEL ensures the provision of universal access under a written contract between a public authority and a private provider where the private operator is obliged to provide services under pre-specified terms and conditions. Unlike VPP, where the provision of universal access is voluntary in nature, the existence of a formal contract under FITEL ensures the achievement of universal access for rural inhabitants.

Part-II: Welfare Implications of the Use of ICTs

5 CONCEPTS OF WELFARE MEASUREMENT

5.1 Introduction

This part of the study deals with the welfare aspects of the use of ICTs. Having described the institutional mechanism of how to provide access to ICTs for rural households in the previous part of the study, it is then necessary to evaluate the welfare impact of the use of ICTs on rural households. Consequently, this chapter and the subsequent four chapters measure the welfare impacts of the use of telecommunications services. While the present chapter outlines the general methodological concepts necessary for the welfare evaluation, the subsequent chapters quantify the magnitude of welfare gain associated with the access to and use of a telephone.

Among the different measures that can be taken to evaluate the impact of the telephone, for instance, efficiency, welfare and sustainability, the current study opts for the analysis of the effects on household's welfare resulting from the change in access to telephones. Literature on the welfare implications of a particular technological change goes back to Griliches's study on the impact of hybrid corn, where he used consumer and producer surplus to measure social welfare.[87] As rural households can be considered both as production and as consumption units, such households can use telephones as a production input as producers and as a consumption good as consumers. Therefore, an ideal measure of welfare impact of telephone use will reflect the change in utility derived from telephone services consumed, given the consumption of all other goods and services, and such a measure will estimate the impact on profit resulting from telephone use in the production process. Instead of joint estimation, the study opts for analysing these two impacts, impact on consumption and impact on production separately under direct and indirect impacts, respectively. More specifically, it asks: what is the direct welfare impact due to change in access, and what is the indirect welfare impact due to use of a telephone.

Direct Impacts: Access of rural households to telecommunication services is important to the well-being of the households as such access reflects a household's welfare standing and determines a household's welfare as well. Households consume telecommunications services and such services are valued in their own right as households derive utility from the consumption of such services, given the availability, as opposed to traditional alternatives. This is due to connectivity, and information transmission through telecommunications (and ICTs in general) is cheaper and faster than other alternatives. A telephone is useful irrespective of whether the poor or rich people use it. In addition, as poor people are usually poor in terms of human capital and alternatives as well,

[87] See Griliches (1957).

the poor often pay a higher price for information and connectivity compared to the price that they would pay if access to telecommunications were available.

Measuring the direct impacts of improvements in telecommunications involves estimating the magnitudes of direct economic gains associated with changes in telecommunications services. This requires the measurement of benefits to users from the improvements in access to telecommunications services by estimating the underlying (derived) demand for the services. It implies that to measure the households' valuation of the telecommunications services and the underlying utility associated with such services, one needs to measure households' demands for the telecommunications services. To this end, the current study measures the households' demands for telecommunications services, and assigns an economic value; it also estimates the compensating variation, and equivalent variation.

Indirect Impacts: While in direct impact measurement the study considers rural households as consumers of telecommunications services and concentrates on measuring the change in households' utility, in the case of indirect impact measurement it considers rural households as a production unit and the logical step would be to measure the impact of telecommunications services on their profits. However, instead of estimating profit, which would again be a direct measure of the impact of telecommunications, the study concentrates on the impact of telecommunications on households' product and factor market participation.

In developing countries both product market and factor market are characterized with imperfect and asymmetric information, while absence of market is widespread. A seller searching for a buyer is unlikely to be fully informed about all the potential buyers and vice versa, and negotiation costs are usually significant as well. As a result, the extent of market failures and self-sufficiency in production and consumption is a widely observed phenomenon. One important reason for the market failures and the observed self-sufficiency, particularly in agriculture, is the high transaction costs.[88] As information cost is an important part of transaction costs,[89] use of telecommunications in the transaction process can reduce transaction costs, overcome problems related to information asymmetries and help market emergence. In the case of existing markets, use of telecommunications may result in increased activities.

5.2 Measuring the Direct Impacts

Assume that a household that derives utility from telephone consumption and consumption of all other goods given by:

---

[88] See for example, Key (2000). See also Stiglitz (1988) for an information theoretic based explanation on market failure.

[89] See Dasgupta and Stiglitz (1988)

(5.1) $u = u(x, z, s)$

where x is the amount of telephones consumed, z is the vector of all other goods that the household consumed, and s is the vector of household characteristics and other shifters.[90] If p is the price of a telephone, q is the price (vector) of z, and y is the income, the demand for x can be expressed as

(5.2) $x = x(p, q, y, s)$

and the indirect utility function that can be derived after constraint maximization is given by

(5.3) $v = v(p, q, y, s)$

Due to the change in availability of a telephone, a household faces a different price for a telephone and it also has a different level of income. So the welfare impact of a change in access to a telephone for the household can be expressed as:

(5.4) $\Delta u = v(p_0, q, y_0, s) - v(p_1, q, y_1, s)$

Here, (p_0, y_0) are the price of a telephone and household income before the change in telephone supply, respectively, and (p_1, y_1) are the price and income after the supply shock. However, the change in welfare represented by (5.4) is ordinal. For a measurable indicator of welfare, a number of monetary measures of welfare are available, for instance: consumer surplus, compensated variation and equivalent variation.[91] They are reviewed in brief as some of them will be applied in the subsequent chapters.

Consumer surplus: Consumer surplus is one of the commonly used measures in evaluating households' welfare effect of a single price change. Consider a change in price of a telephone (x) from p_0 to p_1 with (q, y, s) held constant. The consumer surplus's measure of the effect of this price change, consumer surplus (*CS*), is defined by:

(5.5) $CS = \int_{p_0}^{p_1} x(p, q, y, s)' dp$

Compensating variation: Compensating variation is the maximum amount of money that can be taken away from a household due to the change in telecommunications access status without changing the household's utility level. In other words, compensating variation measures the maximum amount

[90] See Mitchell (1978) for a separable demand function on telephones.

[91] See Sadoulet and Janvry (1995), pp. 11-17.

that a household is willing to pay due to the change (positive) in access and can be expressed as:

$$(5.6) \quad v(p_1,q,y_1-CV,s)=v(p_0,q,y_0,s)$$

Here, CV stands for compensating variation. If $e(p,q,u,s)$ is the expenditure function that shows the minimum income necessary to achieve the level of utility u at given price p, setting both sides of (5.6) equal to u and solving for the expenditure function provides an equivalent expression for compensating variation in terms of the expenditure function:[92]

$$(5.7) \quad CV=e(p_1,q,u_0,s)-e(p_0,q,u_0,s)$$

Equivalent Variation: Equivalent variation is the minimum amount of money that needs to be paid to a household due to the change in telecommunications access without changing the household's level of utility. In other words, the equivalent variation measures the minimum amount that a household requires accepting the (negative) change in access and can be expressed as:

$$(5.8) \quad v(p_1,q,y_1,s)=v(p_0,q,y_0+EV,s)$$

Here, EV stands for equivalent variation. Similar to CV, EV can also be expressed in terms of expenditure function as:

$$(5.9) \quad CV=e(p_1,q,u_1,s)-e(p_0,q,u_1,s)$$

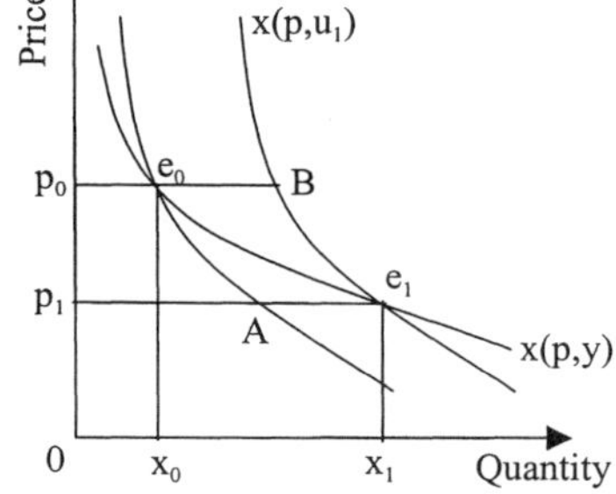

Figure-5.1 Consumer Surplus, Compensating Variation and Equivalent Variation

Figure-5.1 represents consumer surplus and compensating variations together. Here x_0 and x_1 are the past and present provisions (supply) of telephones in rural areas, respectively, and p_0 and p_1 are the associated prices; x(p,y) is the

[92] See LaFrance (1991) for more details on derivation and relations.

Marshallian demand curve, and $x(p,u_0)$ and $x(p,u_1)$ are the Hicksian demand curves. Due to the increased provision of telecommunications, (x_1-x_0) price has decreased from p_0 to p_1. The decrease in price could be due to decrease in tariff or due to decrease in non-tariff costs such as travelling time to the nearest telephone and transport costs or due to both. In these circumstances, the *CV* is the area bounded by $p_0e_0Ap_1$, ΔCS is the area bounded by $p_0e_0e_1p_1$, *EV* is the area bound by $p_0Be_1p_1$.

Both *CV* and *EV* can be applied to evaluate the impact of the telecommunications project. For instance, if the total sum of willingness to pay and compensation requirements as computed with CV is positive, it suggests that a proper compensating scheme could make the project acceptable to everyone and vice versa. In that respect, *CV* provides the basis for the design of compensating schemes that makes everybody at least as well off after the implementation of a scheme as before, which is the definition of Pareto optimality.[93] Similarly, in the case where the sum of *EV* is positive, the telephone project is acceptable to everyone.

5.3 Measuring the Indirect Impacts

For the indirect impacts, that is the impact of a telephone on a household as a producing entity, it starts with a household whose profit function is given by:

(5.10) $$\pi = \pi(z,q,x,p,l,w)$$

Where z is the vector of goods produced by the household and q is the corresponding prices; x is the use of a telephone in production (including marketing) and p is the corresponding price; l is the vector of all other inputs used in the production of z and w is the vector of corresponding prices. If the production function of the household is given by:

(5.11) $$f = f(z,x,l;s)$$

where s is the vector of production shifters, the supply function and factor demand functions of the household can be derived after constraint maximizations are given by:

(5.12) $$z = z(q,p,w;s)$$
(5.13) $$x = s(q,p,w;s)$$
(5.14) $$l = l(q,p,w;s)$$

Here (5.12) is the output supply function, and (5.13) and (5.14) are the factor demand functions.

[93] See Just et al (1980) and Sadoulet et al (1995).

Instead of measuring the impact of telephone use on profit, the study is interested in the households' supply response in the form of market participation in the presence of transaction costs. Figure-5.2 shows the supply response of households in the presence of transaction costs.[94] Here t is the transaction cost that a household incurs as a buyer of a product in addition to market price q, and as a seller of the product a household receives the price net of t. As a result, when the effective price net of transaction cost falls below $q-t$, the household participates in the market as a buyer, and when the effective price is above $q+t$, the household participates as a seller in the market, and in between the price band, $q-t \leq q \leq q+t$, the household remains in autarky. In the figure, ABCD is the whole supply curve, where AB is the buyer's supply curve, CD is the seller's supply curve, and in between these two shows the supply curve of an autarkic household. In the presence of fixed transaction costs, participation in the market is further delayed both for buyers and for sellers.

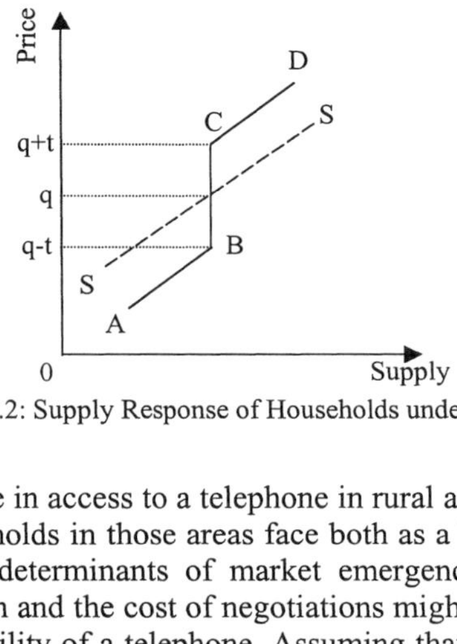

Figure-5.2: Supply Response of Households under the Presence of Transaction Costs

The change in access to a telephone in rural areas may reduce transactions costs that households in those areas face both as a buyer and as a seller. Some of the important determinants of market emergence such as the cost of acquiring information and the cost of negotiations might be reduced due to the increase in the availability of a telephone. Assuming that a telephone has an impact on the reduction of transaction costs, such impact should be reflected in the rural households' supply response, which will be analysed under the indirect impact. However, due to the lack of necessary data in the case of Peru, the analysis of indirect impact will remain limited to the households of Bangladesh only.

[94] This figure is adopted from Key et al (2000), p. 250.

6 MEASURING THE DIRECT IMPACTS: COMPENSATING VARIATION

6.1 Introduction

This chapter and the subsequent three chapters measure the welfare impacts of public telephone. This chapter measures the welfare implications of public telephones in rural areas through the measurement of Hicksian compensating variation, henceforth CV. CV measures the maximum amount of money that can be taken away from a household due to the increased supply of telecommunications leaving the household at its pre-telephone level of utility. In other words, it measures the maximum amount of money that a household would be willing to pay due to the increased access to telecommunications services. Chapter 5 provides a conceptual framework on Hicksian compensating variation, equivalent variation and Marshallian consumer surplus. The advantage of Hicksian compensating variation as a welfare measurement tool over Marshallian consumer surplus is that the former does not suffer from path dependency, for instance, adjustment of price and income, as in the case of the latter. As a result, unlike Marshallian consumer surplus, measurement of Hicksian compensating variation provides a unique monetary measurement of the change in welfare resulting from any intervention.[95]

Access to telecommunications may be explicitly rationed by direct costs such as tariff, or implicitly rationed by indirect costs such as transport cost and opportunity cost of time. In the case of Bangladesh and Peru, households living in rural areas were not explicitly rationed through high and differential tariffs. However, as availability of a public telephone was limited mostly to urban areas until recently, rural households were implicitly rationed through transport cost and opportunity cost of travelling time. Due to the regulatory nature of the telecommunications industry, direct cost of public telephone use did not vary between rural and urban areas depending on the cost of provision, and between poor and rich households depending on the ability to pay. However, despite the absence of variation in direct costs of telephone use, rural households both in Bangladesh and in Peru were rationed by indirect costs of telephone use. Households that were willing to pay the tariff as well as transport cost and opportunity cost of time could travel to the nearest publicly accessible telephone. However, if a household was willing to pay the tariff but not the transport cost and opportunity cost of time it was effectively excluded from the telecommunications services.

Table-6.1a shows the one-way travel time to the nearest public telephone. As expected, in the case of Bangladesh, a rural household's usage rate depends on travel time among other factors; the longer the travel time to the nearest

---

[95] See Sadoulet et al (1995), pp. 12-13.

telephone, the lower the usage. This suggests that the reduction in travel time to the public telephone may increase a household's telephone use. Though for Peru usage rate does not decrease with distance to the public telephone, one-way transport cost to the telephone increases with distance and the average transport cost for the four categories of households described in table-6.1 are S/.0.59, S/.0.89, S/.1.02 and S/.2.66, respectively. In addition, one-way distance to the nearest public telephone and usage rate, and one way transport cost to the nearest telephone and usage rate also show a similar trend in the case of Bangladesh (Table-A6.1a and table-A6.1b in Appendix-III).

Table-6.1: One-Way Travel Time to the Nearest Public Telephone and Usage Rate

One Way Travel Time	Percentage of Sample in Category	Usage Rate (%)
Bangladesh		
Within the village[1/]	15.85	64.44
Within 30 minutes distance	45.07	26.56
Within one hr distance	27.82	21.52
More than one hr distance	11.27	18.75
Total	100.00	30.28
Peru		
Within the village[1/]	81.9	78.00
Within 30 minutes distance	5.76	98.25
Within one hr distance	2.02	100.00
More than one hr distance	10.21	99.02
Total	100.00	81.80

[1/] Zero or negligible distance.

Thus, for the present study, the real question is: what is the direct welfare gain that the rural households in Bangladesh and Peru have attained due to the change in their access to a public telephone? In other words, how much monetary gain is associated with the implementation of the VPP program in Bangladesh and FITEL project in Peru? To answer this question, the present chapter estimates the compensating variation associated with the reduced costs over alternatives to a telephone. The chapter will also look at the distribution of welfare gain associated with the use of a telephone for different income groups.

The estimation of the welfare consequence of a public telephone in rural areas requires the estimation of tariff elasticities and the effects of non-tariff costs such as travel time and costs. However, the estimation is complicated by the fact that, due to the regulatory nature of the telecommunications industry, the tariff usually does not vary within a country. However, to account for this, the present study uses private-time-price variation to identify the necessary parameters.[96] The study also allows a natural interaction between the price and

[96] See Becker (1965) for his seminal work on this issue.

income in demand function. Studies on demand usually assume price effect is independent of income effect, which is restrictive in nature.[97]

The remainder of the chapter proceeds as follows: section 6.2 describes the theoretical framework, section 6.3 describes the empirical estimation where conditional utility functions and demand functions are described along with estimation results and compensating variation. This section is followed by conclusions.

6.2 Theoretical Framework

Households can use telecommunications services both as consumption good and as production input as described in chapter 5. The services that households derive from the use of a telephone are important for the well-being of the households as such services reflect the households' welfare standing and determine the households' welfare. The fact that not all households are users of telephones given the access reflects that the alternatives to telecommunications offer better services for such households. In addition, households' observed and unobserved characteristics may also play a role.

Following Mitchell (1978), it is assumed that each household has a separable utility function that depends on the consumption of telecommunications services and consumption of all other goods. The consumption of telecommunications services is discrete; at any given time the household decides to use a telephone or not. Given the fact that there is hardly any provision at household level, a decision for telephone consumption means deciding on using a public telephone, which includes direct cost such as tariff, and indirect costs such as value of travel time and transport cost. Households compare the utility derived from the use of a telephone with the traditional alternatives and choose the option that provides them with the highest utility.

So the expected utility for household i conditional on telephone use is given by:

$$(6.1) \qquad U_i^1 = U(T_i, Z_i)$$

where T_i is the ith household's consumption of telecommunications services, and Z_i is the consumption of all other goods net of telecommunications services. In the case that the household does not consume telecommunication services, the utility for the household i is given by:

$$(6.2) \qquad U_i^0 = U(C_i)$$

[97] See Gertler et al (1987) for a discussion on this issue.

where C_i denotes the consumption of the household when it does not consume telephone services. If the household's income is denoted by Y_i, the budget constraint that the household faces is given by:

$$(6.3) \qquad C_i = Y_i$$

In the case that the household consumes telecommunications services, the budget constraint is given by:

$$(6.4) \qquad T_i + Z_i = Y_i$$

For the household, the problem is to maximise utility given by:

$$(6.5) \qquad U_i^* = \max(U_i^0, U_i^1)$$

subject to budget constraint given by (6.3) and (6.4). Here, U_i^* is the maximum utility, and U_i^0 and U_i^1 are the conditional utilities specified in (6.1) and (6.2). The solution to the utility maximization problem gives the probability that the household chooses one alternative over another, which can be interpreted as demand function in a discrete choice model. The demand function will be used to solve unconditional utility functions to assess the welfare impact of access to telecommunications in terms of compensating variation.[98]

6.3 Empirical Estimation

The empirical specification is based on Gertler et al (1987). The utility maximization problem described above provides a system of demand equations whose forms are probabilities. The probability that a household uses a telephone equals the probability that the use of a telephone provides the highest utility among the alternatives and vice versa. The functional form of the demand functions depends on the functional form of the conditional utility function and on the distribution of the stochastic terms.

6.3.1 The Conditional Utility Functions

As argued in Gertler et al (1987), income can influence a household's choice of telephone use only if the conditional utility function allows for a non-constant marginal rate of substitution of telephone use for consumption. A parsimonious form of conditional utility function that does not impose the restriction of constant marginal rate of substitution is the semi-quadratic, which is linear in telephone use and quadratic in consumption. It follows that the conditional utility function for the telephone use can be written as:

[98] See Gertler and Glewwe (1989) that estimated compensating variation for education.

(6.6) $$U_i^1 = \beta_1 T_i + \beta_2 Z_i + \beta_3 Z_i^2 + \varepsilon_i^1$$

Here, ε_i is a randomly distributed error terms with zero mean and unit variance. Consumption net of telecommunication expenditure can be derived from budget constraint and can be given by:

(6.7) $$Z_i = Y_i - P_i$$

where P_i is the cost of telephone use that includes both direct cost such as tariff and indirect cost such as transport cost and value of travel time. Substituting (6.7) into (6.6) gives the utility derived from the consumption of telephone and the consumption of other goods:

(6.8) $$U_i^1 = \beta_1 T_i + \beta_2 (Y_i - P_i) + \beta_3 (Y_i - P_i)^2 + \varepsilon_i^1$$

In the case that the household does not use a telephone, the utility is given by:

(6.9) $$U_i^0 = \beta_2 Y_i + \beta_3 Y_i^2 + \varepsilon_i^0$$

The identification of the parameters in (6.8) and (6.9) requires that the value of consumption of a telephone and consumption of non-telephone goods vary across alternatives and it is the variation in these values across alternatives that allow for the identification of the parameters. As shown in Gertler et al (1987), the transformation of conditional utility functions reveals that it is the price-income interaction term that allows the income to have influence on alternatives.

To allow for the household's observed characteristics, it is assumed that the utility from the consumption of a telephone can be expressed by:

(6.10) $$\beta_1 T_i = \gamma_j X_i^j + \eta_i$$

where X_i^j is a vector of a household's demographic and other characteristics and η_i is the error term. Substituting (6.10) into (6.8), the conditional utility function for the households that consume telecommunications services can be rewritten as:

(6.11) $$U_i^1 = V_i^1 + \varepsilon_i^1 + \eta_i^1$$

where

$$V_i^1 = \gamma_j X_i^j + \beta_2 (Y_i - P_i) + \beta_3 (Y_i - P_i)^2$$

Equivalently the conditional utility for the households that do not use a telephone given by (6.9) can be rewritten as:

(6.12) $$U_i^0 = V_i^0 + \varepsilon_i^0$$

where

$$V_i^0 = \beta_2 Y_i + \beta_3 Y_i^2$$

6.3.2 The Demand Functions and Welfare

The demand for a telephone, that is the probability that a household will use a telephone, is given by:[99]

(6.13) $$\pi^1 = \frac{\exp(V^1)}{\exp(V^0) + \exp(V^1)}$$

In the case that the household does not use a telephone, the demand function is given by:

(6.14) $$\pi^0 = (1 - \pi^1)$$

The demand functions given by (6.13) and (6.14) can be used to measure the household's willingness to pay for the provision of a public telephone closer to their settlements. Here, the willingness to pay is the compensating variation that keeps the utility of the households at its previous level, which is welfare neutral by definition.

Following Small and Rosen (1981), the compensating variation in the case of the logit model can be measured by:

(6.15) $$CV = (1/\lambda)\{\ln[\exp(V^0) + \exp(V^1)] - \ln[\exp(V^0) + \exp(V_k^1)]\}$$

where CV is evaluated at four different alternatives denoted by *k*.

6.3.3 Data and Summary Statistics

Similar to other chapters, the present chapter also utilizes data from the household surveys conducted in six rural regions in Bangladesh and the south rural region in Peru. The data collection procedures and methods have been described in appendix-I. For both countries, surveys followed a two stage random sampling procedure and focused on the respondent's use of the public telephone service, and direct and indirect costs associated with the use. In addition the surveys also contained questions on households' demographic and housing characteristics, employment, income and expenditure, participation in social programs and availability of infrastructure, among other factors. Both

[99] The subscript i has been avoided here.

surveys included settlements with and without public telephones where a lack of a public telephone service was primarily due to a supply constraint instead of one of demand.[100]

Table-6.3.3 presents summary statistics of the household's use status; annual household expenditure; education, occupation and age of the head of the households; size of the households; and village public telephone dummy. Summary statistics for the full set of variables that are included in the estimation are provided in appendix-III in table A.6.3.3a and table-A.6.3.3b. Among the covariates included in the estimation, household expenditure is measured in local currencies, age in years, and user-status and occupation are dummy variables. The other covariate that has been included in case of Bangladesh is the gender of the head of the household. In the case of Peru, additional covariates are the benefits of a public telephone, a household's media use habit, and a native language dummy.

Table-6.3.3: Summary Statistics

Covariates	Bangladesh		Peru	
	Mean	Std. Dev	Mean	Std. Dev
User Status (user=1, nonuser=0)	0.30	0.46	0.82	0.39
Annual Household Expenditure (in US$)	1455.29	1753.40	2415.30	2093.31
Years of Edu. of the Household Head[1/]	0.84	0.37	7.49	4.66
Occupation of the Household[2/]	0.54	0.50	0.24	0.43
Age of the Household Head	46.57	12.74	44.26	14.05
Family Size	5.38	1.74	4.90	2.03
Village has a Telephone (yes=1)	0.5	0.50	0.5	0.5

[1/] In the case of Bangladesh it is a literate-illiterate dummy; [2/]In the case of Bangladesh, it is an agriculture-non-agriculture dummy and in the case of Peru, it is whether the household has a room for economic activity or not. For Bangladesh figures are for 2001, and for Peru figures are for 1999. The exchange rate for Bangladesh was 1US$=Tk.56. For Peru, the exchange rate was 1US$=S/.3.38. Exchange rate source for Peru: WDI 2001.

Direct and Indirect Costs of Telephone Use: For the estimation purpose and for the calculation of CVs, both direct and indirect costs of public telephone use have been taken into account. In the case of Bangladesh, the direct cost of telephone use includes the total cost of each telephone call that a household has made, two-ways transport cost to the nearest public telephone, and opportunity cost of time of a return journey converted into the prevailing wage rate. In the case of Peru, the total cost of a telephone includes all the components considered in the case of Bangladesh. In addition, the total cost of public telephone use also includes the cost of food and accommodation cost that some households needed to spend based on the distance to the nearest public telephone.

[100] Many of the settlement inhabitants without access to a public telephone service travel many kilometres to arrive at a settlement that has the service.

Alternatives to public telephone: Both in the case of Bangladesh and of Peru, four means of communication are considered that households would use as alternatives to a public telephone. The users households have stated those alternatives against each telephone call that the respective households have generated from the recently available public telephone. As a result, for a single household, alternatives to a telephone may vary. However, for each telephone call, there is only one alternative. For Bangladesh, the stated alternatives to a public telephone are to telephone from other places, for instance travelling to the nearest town; sending a letter from the nearest post office; travelling to the recipient, or sending a person there; and others such as a combination of travelling and making a telephone call from other places. For Peru, the stated alternatives are sending a letter from a mail office; sending a letter using transport agencies; communicating through radio; travelling to the recipient, or sending a person there.

6.3.4 Estimation Results

The estimation of compensating variation involves two stages that need to be differentiated: first, the parameter estimation stage, and second, the compensating variation calculation stage. Therefore, in order to calculate compensating variation, a logit model of telephone use has been estimated with three different specifications. The different specifications are assumed to capture the stability of estimated coefficients. Table-6.3.4a presents the logit coefficients and table-6.3.4b presents the standardized logit coefficients for Bangladesh, and table-6.3.4c and table-6.3.4d present the corresponding coefficients for Peru. As can be seen from table-6.3.4a and table-6.3.4b, the estimated coefficients have remained stable and consistent across different specifications. For the calculation of CV, it is only the full specification that has been used (labelled as model 3).

Table-6.3.4a: Parameters Estimate for Telephone Use, Method: Logit, Bangladesh
Dependent Variable: Household use telephone or not (yes=1, no=0)

Covariates	**Model 1**		**Model 2**		**Model 3**	
Consumption ($x10^{-2}$)	0.021		0.02		0.02	
	(0.004)	**	(0.004)	**	(0.004)	**
Consumption Squared ($x10^{-4}$)	-1.92e-05		-1.58e-05		-1.78e-05	
	(5.79e-06)	**	(6.09e-06)	**	(6.21e-06)	**
Household Characteristics						
Education of the head of the HH			1.6185		1.6556	
			(0.6631)	**	(0.6601)	*
Occupation of the head			-0.6333		-0.7062	
			(0.3173)	*	(0.3228)	*
Age of the head of the HH			0.0067		-0.0397	
			(0.0748)		(0.0793)	
Age squared			0.00001		0.0004	
			(0.0007)		(0.0008)	
Size of the HH			0.1014		0.1019	

			(0.0961)		(0.0986)	
Sex of the head male=1)			-1.4678		-1.4249	
			(0.6306)	*	(0.6458)	*
Village has a Public Tel (yes=1)					0.8831	
					(0.3250)	**
Constant	-2.1075		-2.555		-1.9209	
	(0.2593)	**	(1.9713)		(2.0237)	
Log likelihood	-145.95		-136.69		-132.84	
Number of observations	280		280		280	
LR chi2	46.72		65.25		72.95	
Probability > chi2	0.0000		0.0000		0.0000	
Pseudo R2	0.1380		0.1927		0.2154	

Standard errors are in parentheses; * significant at 5% level; ** significant at 1% level

Table-6.3.4b: Parameters Estimate for Telephone Use, Standardized Coefficients Method: Logit, Bangladesh

Dependent Variable: Household use telephone or not (yes=1, no=0)

Covariates	**Model 1**		**Model 2**		**Model 3**	
Consumption (x10^{-2})	0.00416		0.00318		0.00341	
	(0.001)	**	(0.001)	**	(0.001)	**
Consumption Squared (x10^{-4})	-3.86e-6		-2.96e-6		-3.29e-6	
	(0.0)	**	(0.0)	**	(0.0)**	
Household Characteristics						
Education of the head of the HH			0.2235		0.2236273	
			(0.0581)	**	(0.05645)	*
Occupation of the head of the HH			-0.1198		-0.1321879	
			(0.06)	*	(0.0607)	*
Age of the head of the HH			0.0012		-0.0073519	
			(0.014)		(0.01468)	
Age square of the head of the HH			2.46e-06		.0000786	
			(0.00014)		(0.00014)	
Size of the HH			0.01899		0.0188562	
			(0.01795)		(0.01821)	
Sex of the head (male=1)			-0.3362		-.3244714	
			(0.151)	*	(0.15588)	*
Village has a public telephone (1)					.1633277	
					(0.05956)	**
Log likelihood	-145.95		-136.69		-132.84	
Number of observations	280		280		280	
LR chi2	46.72		65.25		72.95	
Probability > chi2	0.0000		0.0000		0.0000	
Pseudo R2	0.1380		0.1927		0.2154	
Predicted Probability	0.2486		0.2496		0.2452	

Standard errors are in parentheses; * significant at 5% level; ** significant at 1% level

Table-6.3.4c: Parameters Estimate for Telephone Use, Method: Logit, Peru
Dependent Variable: Use of Public Telephone (yes=1, no=0)

Covariates	Model 1		Model 2		Model 3	
Consumption (x 10^{-2})	0.1144		0.1036		0.10140	
	(0.0328)	**	(0.0323)	**	(0.0347)	**
Consumption square (x 10^{-4})	- 0.0018		-0.0016		-0.00154	
	(0.0005)	**	(0.0005)	**	(0.0005)	**
Years of Edu of the Head of the HH	0.1359		0.13		0.12944	
	(0.0260)	**	(0.0265)	**	(0.0279)	**
Room used exclusively for economic	0.8257		0.7651		0.76060	
activity (Yes=1, No=0)	(0.2383)	**	(0.2401)	**	(0.2437)	**
Age of the Head of the HH	0.0304		0.0322		0.02585	
	(0.0356)		(0.0356)		(0.0389)	
Age of Square of the Head of the HH	-0.0003		-0.0004		-0.00033	
	(0.0004)		(0.0004)		(0.0004)	
Native Language (Yes=1, No=0)	-0.7351		-0.7386		-0.66932	
	(0.2333)	**	(0.2355)	**	(0.2411)	**
Use of Media (No. of activities)			0.0207		0.02769	
			(0.0166)		(0.0167)	~
Village with telephone (Yes=1, No=0)			0.4014		0.43518	
			(0.1695)	**	(0.1764)	**
Household Size (# of members)					-0.01615	
					(0.0466)	
Identified Benefits from the use						
- Fast communication (Yes=1)					0.51609	
					(0.1952)	**
- Avoid trips (Yes=1)					0.66407	
					(0.2344)	**
- Avoid isolation (Yes=1)					1.01903	
					(0.3261)	**
- Save money (Yes=1)					0.55879	~
					(0.3269)	
- Save time (Yes=1)					0.07738	
					(0.3042)	
Constant	-0.3902		-0.6942		-1.13186	
	(0.9176)		(0.9190)		(0.9536)	
No of observations	991		990		987	
Log Likelihood	-452.4		-448.2		-434.7	
Wald chi2	121.02		133.47		148.6	
Probability > chi2	0.000		0.000		0.000	
Pseudo R2	0.141		0.148		0.171	

Standard errors are in parentheses; ~significant a 10% level; * significant at 5% level; ** significant at 1% level

Table-6.3.4d: Parameters Estimate for Telephone Use, Standardized Coefficients, Method: Logit, Peru, *Dependent Variable: Use of Public Telephone (yes=1, no=0)*

Covariates	Model 1		Model 2		Model 3	
Consumption (x 10^{-2})	0.0172		0.0156		0.0148	
	(0.0046)	**	(0.0046)	**	(0.0047)	**

Consumption square (x 10^{-4})	-0.0003		-0.0002		-0.0002	
	(0.0001)	**	(0.0001)	**	(0.0001)	**
Years of Edu of the Head of the HH	0.0198		0.0188		0.0178	
	(0.0036)	**	(0.0037)	**	(0.0037)	**
Room used exclusively for economic	0.108		0.1003		0.0959	
activity (Yes=1, No=0)	(0.0446)	**	(0.0445)	**	(0.0262)	**
Age of the Head of the Household	0.0043		0.0046		0.0032	
	(0.0054)		(0.0054)		(0.0056)	
Age square of the Head of the HH	-0.0001		-0.0001		0.0000	
	(0.0001)		(0.0001)		(0.0001)	
Native Language (Yes=1)	-0.1024		-0.1026		-0.898	
	(0.0461)	**	(0.0490)	**	(0.0278)	**
Use of Media (No. of activities)			0.003		0.0040	
			(0.0024)		(0.0024)	~
Village with telephone (Yes=1)			0.0651		0.0699	
			(0.0376)	**	(0.0260)	**
Household Size (# of members)					-0.0020	
					(0.0067)	
Identified Benefits from the use:						
- Fast communication (Yes=1)					0.0771	
					(0.0293)	**
- Avoid trips (Yes=1)					0.0870	
					(0.0263)	**
- Avoid isolation (Yes=1)					0.1133	
					(0.0269)	**
- Save money (Yes=1)					0.0695	~
					(0.0344)	
- Save time (Yes=1)					0.0029	
					(0.0408)	
No of observations	991		990		987	
Log Likelihood	-452.4		-448.2		-434.7	
Wald chi2	121.02		133.47		148.6	
Probability > chi2	0.000		0.000		0.000	
Pseudo R2	0.141		0.148		0.171	
Hit Rate	0.769		0.776		0.784	

Standard errors are in parentheses; ~significant a 10% level; * significant at 5% level; ** significant at 1% level

Turning to the individual coefficients, the coefficients of consumption and consumption squared are significant at one percent level, which implies that marginal utility gain from consumption (net of telephone) diminishes relative to the gain from the consumption of telecommunication services. Direct and indirect costs of telephone use are incorporated into the consumption variables and it is the variation in prices that is significant here. The coefficient of a village dummy that expresses whether a village has a public telephone or not is significant and positive implying that the availability of a telephone at a minimum distance significantly increases telephone consumption. Other covariates that are significant are education and occupation of the head of the household. Literate household heads are more likely to be telephone users, so too the non-agricultural households. In Bangladesh, female household heads are more likely to use a telephone than their male counterparts; and for Peru, use of the media has a negative effect and benefits of public telephone dummies have positive significant effects.

6.3.5 Compensating Variation

As shown in chapter 3, for rural households, access to a telephone is implicitly rationed by travel time and cost of transport both in Bangladesh and in Peru. Under a regulated tariff structure that takes care of direct cost, one way to increase the access is through reducing indirect costs by making provision for public telephones in each settlement. The recent expansion of telephones through the VPP program in Bangladesh and the FITEL project in Peru, described in chapter 4, has increased the access of rural households. To evaluate the direct welfare benefits of the project, the compensating variation has been estimated here.

Table-6.3.5: Estimated Compensating Variation: Alternatives to the Service of Rural Telephony[1/]

	Quartile of Expenditure			
	Bottom 25% I	II	III	Top 25% IV
Alternatives				
	Compensating Variation per Call (US$)			
Bangladesh				
Telephone from other place	0.15	0.21	0.20	0.20
Sending letter (post office)	0.11	0.14	0.13	0.13
Travel	0.65	0.84	0.78	0.81
Others	1.19	1.59	1.44	1.56
Peru				
Sending Letter (mail office)	2.01	1.93	1.81	1.72
Sending letter (transport agency)	1.62	1.57	1.51	1.47
Radio	1.67	1.58	1.52	1.45
Travel	2.91	2.74	2.52	2.07
Cost of Telecommunications				
	Direct + Indirect Costs per Call (US$)			
Bangladesh	0.98	0.54	0.76	0.88
Peru	0.86	0.93	1.13	1.33

[1/]Compensated Variation following Small and Rosen (1981); Exchange Rate 1US$=Tk.56 (2001); 1US$=S/.3.38 (1999)

Whether households are benefiting from the increased access to a telephone depends on whether the gain in welfare from having a telephone closer is higher than the direct and indirect cost of the use of a telephone. The compensating variation is the maximum amount of money that the households are willing to spend on a telephone that will keep their welfare unchanged (welfare neutral). Compensating variations of the telephone have been estimated using formula given by (6.15).

Table-6.3.5 shows the estimated compensating variations (CV). CVs have been estimated over four alternatives to the service of a rural telephone. The alternatives that rural households in the two study countries use vary to a certain extent. Nonetheless, they are comparable as well. The estimated compensating variations are the benefits over the alternatives that the rural households had to use in the absence of recently installed rural telephones in both countries. The CV varies across different income groups and among different alternatives to rural telephones. As a result, CVs are presented in table-6.35 for different income groups under different alternatives. To facilitate comparison, the local prices are converted into US$ based on the nominal exchange rate which prevailed during the time of the surveys.

The estimated CVs show that, for all income groups and for both countries, the rural telephone projects are welfare improving over the alternatives of rural telephones in general, and use of a telephone is directly contributing to rural households' welfare in particular. In the case of Peru, the estimated CV per call is higher for the bottom income quartile than the top income quartile under all four alternatives. This implies both an absolute and a relative welfare gain from the rural telephony project in Peru. In the case of Bangladesh, though the bottom income quartile does not receive higher benefits than the top income quartile, the absolute welfare gain from the use of a telephone as a part of income is higher for this group than the other groups.

The total cost of a telephone per call that includes both direct and indirect costs also shows a positive relationship with income in Peru. However, it does not have any clear relation with income in Bangladesh. In fact, the households of lowest quartile of income incur the highest cost of telephone per call and this in fact explains the low CVs of this group. In contrast to Bangladesh, CVs are higher in Peru. However, the costs of telephone calls in Bangladesh are within the range of the costs of telephone calls in Peru. This is again an evidence of the relative scarcity of access to telephones in rural areas, which is starker in Bangladesh compared to Peru.

6.4 Conclusions

The estimation of the welfare benefits of a public telecommunications project is complicated by the fact that due to the regulatory nature of the telecommunications industry, telephone tariffs does not vary across the country. However, under such circumstances, the variation in an individual's private-time-price can be used to identify the parameters necessary to measure the welfare of such projects.

This chapter has measured the welfare implications of rural telecommunications projects in two countries, Bangladesh and Peru, through the measurement of Hicksian compensating variation. The advantage of Hicksian compensating variation is that it offers the measurement of welfare neutral prices that keep the households' utility level unchanged. However, as CV varies for different income groups and for different alternatives to telephones, to observe the welfare measurements for different groups, the measurement needs to shows the distribution according to income quartile and according to alternatives.

The present study has measured the Hicksian compensating variation under different alternatives for four different income quartiles. The estimated compensating variations show that the rural telephony projects, VPP program in Bangladesh, and FITEL project in Peru, are welfare enhancing in general and they enhance distributional equity in some cases.

7 MEASURING THE DIRECT IMPACTS: WILLINGNESS TO PAY FOR THE RURAL TELEPHONE SERVICES

7.1 Introduction

This chapter measures the direct welfare impact of telephones on rural households by measuring the equivalent variation. While in the previous chapter welfare gain associated with the use of a telephone is measured without changing the household's utility level, in the present chapter welfare gain is measured based on the household's new utility level caused by the availability of a telephone (real and hypothetical). For this, it measures the rural households' willingness to pay for access to public telephone services in Bangladesh and Peru through contingent valuation methods. The development of contingent valuation methods together with the econometric expansion of qualitative response models has permitted an approximation to the consumer surplus (Hicksian equivalent variation) in the presence of externalities, public good and information asymmetries. This approximation can be made by the simulation of hypothetical markets that allow the seizure of what will be called the willingness to pay (WTP).[101]

Through the recreation of hypothetical markets, information about consumer preferences can be obtained from free-response questions or using bounded-response questions. The first choice has not been commonly used in contingent valuation literature.[102] On the other hand, there is a consensus in the usage of the second choice in contingent valuation surveys. This second choice can be classified in a simple referendum where a simple offer is made, or in a double referendum where a second offer is made conditioned to the response given to the first offer.

One of the most popular survey designs to select the willingness to pay is the dichotomous choice (simple referendum). In this particular case, respondents to the survey are asked whether they are determined to pay a certain amount of money. The question is simple and easy to answer. However, it supplies a limited quantity of information. In this perspective, a sequential referendum would increase the valuation efficiency but at the expense of a greater difficulty in the estimation of the distribution.[103]

---

[101] See Hanemann (1984); McFadden (1994); Maddala (1983). An extensive literature has investigated the use of the contingent valuation method to value the environmental good. For instance, see Bowker and Stoll (1988), Cameron and Huppert (1991), Cameron and James (1987) and the references in McFadden (1994).

[102] The option of open response has relatively a high incidence of no answer or "protest" answer.

[103] See Hanemann (1984); McFadden (1994); Haab and McConell (1997)

The analysis frequently observed in the literature of contingent valuation emphasizes the conditional analysis of the willingness to pay, where one of the main interests is to analyse the conditional relation of willingness to pay and the vector of covariants (X), which describes the consumer characteristics. McFadden (1994) shows that, when necessary, auxiliary information is available to re-weight a sample then there is a gain in statistical efficiency of a conditional approximation. This result does not require the covariants to be free of measurement errors or the conditional mean of the WTP to be linear in X, because errors simply contribute to the non-conditional variance. However, he warns that the conditional approximation is not robust to inconsistencies between sample data and auxiliary distributions of covariants, such as mean differences due to changes in the phraseology of the question, codification or the period of data collection.

The estimation of the distribution of the WTP is tackled in the literature through parametric or non-parametric approximations.[104] The parametric approximation, widely used in the contingent valuation analysis, specifies a parametric model in which a relation between the WTP and the consumer intrinsic characteristics is stated. The non-parametric approximation estimates the distribution of the WTP without assuming any parametric specification of the preferences distribution. Both approximations have advantages and disadvantages. The parametric approximation is susceptible of misspecification errors, being the estimation sensible to the parametric family that contains the distribution of the WTP. The parametric distributions family considered frequently in literature includes logistic, loglogistic, normal, lognormal, bounded normal, bounded logistic and Weibull distributions. On the other side, the non-parametric approximation does not permit a conditional analysis.[105]

The main advantages of the parametric method are that it makes it relatively easy to impose preferences axioms, combines experiments and, primarily, allows one to extrapolate the calculation to different populations without constraining exclusively to the sampled population. On the other hand, the main limitation is that if the parametratization is not flexible enough to describe the behaviour, then the incorrect specification might give inconsistent estimations of the real WTP. Specifically, the estimated mean is highly sensible to the assumption made about the parametric family, which contains the distribution of the WTP.[106]

In the development of referendum models there are two stages that need to be differentiated: first, the parameter estimation stage, and second, the mean

---

[104] See Hanemann and Kanninen (1999) for a comprehensive review of the statistical methods.
[105] See An (2000) on this point.
[106] See McFadden (1994).

willingness to pay calculation stage. Despite the wide diffusion of referendum models, some problems persist in the calculation of the measures of WTP from the estimated parameters. One of the problems arises when the mean willingness to pay is negative. In this case, one possible parametric solution consists in the assumption of a distribution defined strictly for positive values (Lognormal or Weibull) or the assumption of a distribution defined for positive and negative values and the bounding of the estimation of the expected willingness to pay (Bounded Normal or Bounded Logistic). Another possible solution, non-parametric in this case, implies the use of an empirical distribution function based in the proportions of negative responses to each specified offer (Turnbull Approximation).

In this perspective Haab and McConell (1997) explore these tentative solutions and find that models with bounded distributions or distributions defined strictly for positive values tend to produce lower estimations of the mean of the WTP. This fact by itself is not a problem. However, when the range of the willingness to pay that varies between zero and infinite is considered, the estimated means are biased toward higher values.[107] On the other hand, the use of the Turnbull approximation does not permit a simple estimation of the covariants effects over the mean willingness to pay. Therefore, if the objective of the study is to measure or contrast some effects of the characteristics of consumers over the mean willingness to pay, the parametric approximation is convenient. In comparison, if the objective is the estimation of the mean of the willingness to pay, the non-parametric approximation is adequate.

The objective of the present chapter is to estimate the mean willingness to pay of rural households from the two countries for the use of a public telephone service. For Bangladesh, the estimation includes a local telephone service, a long-distance national telephone service, and an international telephone service and for Peru, the estimation includes a local telephone service and an LDN telephone service. The study has not considered international calls in the case of Peru as rural households are primarily the recipients of international calls there and they seldom generate such calls. Apart from estimating WTP, the study will try to estimate the effects of individual characteristics on the mean willingness to pay.

The study is important in considering the multiple benefits that rural households obtain with the access to a public telephone service both in Bangladesh and in Peru. Among the main benefits that have been identified in the survey are: contact with relatives, possibility of purchases or sales, information of prices of agriculture products, knowledge of the market situation before the delivery of

[107] In respect to this last point, recent research has incorporated an upper bound to the willingness to pay defined by the income of the person.

the product to the city, contact with carriers, attention of health emergencies, etc. In effect, information technology access generates externalities that might dynamize the economy of rural households, thus reducing poverty levels and increasing welfare.

The present study will use both parametric and non-parametric approximations that are observed in contingent valuation literature for the distribution and estimation of the WTP. In the development of the parametric approximation it will make use of two alternatives. First, it will estimate a logit model, and with the resulting parameters it will proceed with the estimation of different central tendency measures of the willingness to pay. Second, it will estimate a bounded probit model that includes limits to the willingness to pay, and with the resulting parameters it will estimate the central tendency measures of willingness to pay. For non-parametric estimation, it will estimate WTP willingness to pay based on two methods: Turnbull algorithm, and Kriström's nonparametric method. Once estimated, these central tendency measures, parametric and nonparametric, may be used in the profit/cost analysis of the rural telephone project.[108]

7.2 Modelling the Willingness to Pay

7.2.1 Theoretical Model

It is assumed that the households interviewed answer the contingent valuation questions based on the utility that it represents for them from consumption of rural telephone services. The household utility is defined as a function that depends on the consumption of a compound good Z and on the consumption of rural telephone services, TR, given by:

$$(7.1)\quad U = U(Z,TR) \qquad U_Z > 0 \qquad \text{and } U_{TR} > 0$$

In this case the utility is increasing in Z and TR. Solving the consumer problem, one can obtain the indirect utility of an individual who consumes a rural telephone service given by:

$$(7.2)\quad V_1 = V_1(Y - WTP|TR > 0, X, \varepsilon)$$

where Y is the household income, WTP is a money amount that makes TR not zero, X is a vector of the household's observable characteristics and ε is a vector of non-observable characteristics. Therefore, the optimal willingness to

[108] The study estimates both parametric and non-parametric measures. The purpose is to avoid the problems stated by Duffield and Patterson (1991); they pointed out that the estimated willingness to pay value might be very influenced by the parametric or non-parametric methodology used.

pay by the household is WTP* which makes the following indirect utility functions equivalent:[109]

(7.3) $V_0(Y|TR=0,X,\varepsilon)=V_1(Y-WTP^*|TR>0,X,\varepsilon)$

Then, assuming that $V_1(\lambda, p\ |TR>0,X,\varepsilon)$ is increasing in λ, there exists an inverse function $\psi(\nu,X,\varepsilon)$ which makes $\psi(V_1(\lambda,p|\ TR>0,X,\varepsilon),\ X,\varepsilon)=\lambda$. Therefore, the willingness to pay can be expressed as:

(7.4) $WTP^* = Y-\Psi(V_0(Y|TR=0,X,\varepsilon),X,\varepsilon)\equiv WTP^*(Y,X,\varepsilon)$

This definition of the willingness to pay has some interesting implications and one of them is that being ε a non-observable variable, WTP is a random variable which has a conditional distribution of the individual characteristics that depends of the distribution followed by ε.

7.2.2 Parametric Approximations

7.2.2.1 Logit Model

The parametric estimation of WTP using a logit model consists of the following stages:

i. Estimation of a logit model,[110] with a dependent variable with a value of 1 if the offer (bid) was accepted and 0 if it was not. The explanatory variables include the value of the bid and socio-economic characteristics of households.

Formally, the idea is that the respondent has to answer whether he is willing to pay for a change in access to the telephone from 0 to TR if it costs $B. If one assumes that the indirect utility function is V(Y,TR.X) where X is a vector of individual characteristics,[111] then the answer would be an affirmation if:

(7.5) $V(Y-B,TR,X)-V(Y,X)\geq 0$

---

[109] The stability of the individual preferences is assumed. This is the fundamental assumption of the welfare theory of Hicks.

[110] In literature, logit and probit are frequently used, but McFadden (1994) shows that there are other possibilities in which the Gamma and Weibull distribution can be used. The logit's advantage is that its closed shape is easier to operate mathematically in comparison with other distributions. Moreover, as mentioned in Maddala (1983), despite that the logit distribution has longer tales than the normal distribution (used in probit), both models produce similar predictions of elasticities and probabilities.

[111] The vector of market prices (P) is not included because it is assumed that they are constant.

and a negative one if to the contrary. In this way, following Cameron and James (1987) and Bowker and Stoll (1988), a logistic model can be used. The probability of an affirmative response to a bid of $B when the consumer has a vector X of explanatory variables is given by:

$$(7.6) \quad P(yes/x) = \frac{e^{\beta'X}}{1+e^{\beta'X}}$$

ii. Once the logit model has been estimated, Hanemann (1984,1989), Ardilla (1993) and Vaughan, et. al (1999) propose some measures of willingness to pay (WTP) for the non-constrained mean, median, and bounded mean which constrains the WTP to positive values (detailed in table-7.1). According to their usage in previous research, all of these measures are equally legitimate. For example, the non-bounded median of the logit estimation has been one of the most popular ones, in spite of its potential problem of negative values for the WTP. The present study considers all of the measures and shows the results for each of them.

Table-7.1 shows the different central tendency measures for the probability model. Letter C in Table 7.1 is an abbreviation to identify the measure of the central tendency of the WTP, following the notation used by Hanemann (1984,1989). The term is called augmented intercept and is equivalent to the intercept coefficient of the model plus the other estimated parameters (i=1…n) - except the bid parameter (β)-, multiplied by the sample mean of the explanatory variables ($\overline{x}_i$).

Table-7.1: Central Tendency Measure Formulas for the Logit Model

Description	Symbol	Formula
Mean, E(WTP), -8<WTP<8	C^+	α/β
Median WTP	C^*	α/β
Truncated Mean, E(WTP), 0<WTP<8	C'	$\ln(1+\exp(\alpha))/\beta$
Truncated Mean, E(WTP), $0<WTP<B_{max}$	C^-	$1/\beta \ln[(1+\exp(\alpha))/(1+\exp(\alpha-\beta B_{max}))]$

Note: B_{max} is the maximum bid; The measures with bid logarithmic transformation were not applied because the results were not consistent with the selected bid ranks.

7.2.2.2 Bounded Probit Model

Following Haab and McConell (1997), it is assumed that the willingness to pay for rural telephone services comprises a range that varies from zero to an amount, A, which might be the total income or a fraction of it. Then, assuming that $\varepsilon \sim N(0,\sigma^2)$, WTP* is defined as a stochastic function of the vector of characteristics which considers constraints imposed beforehand. This can be written as:

$$(7.7)\quad WTP = \frac{A_i}{1 + e^{X_i\beta+\varepsilon}}$$

Considering that the respondents interviewed answered the question "...are you willing to pay the amount of t for the rural public telephone service?", one can model the probability response of each respondent as:

$$(7.8)\quad P(WTP_i^* \le t) = P\left(\frac{A_i}{1 + e^{X_i\beta+\varepsilon_i}} \le t\right) = P\left(\frac{\varepsilon_i}{\sigma} \le \frac{X_i\beta - \ln\left(\frac{A_i - t}{t}\right)}{\sigma}\right)$$

Next, the logarithmic likelihood function for the case of a probit model is:

$$(7.9)\quad \ln(L) = \sum_{i=1}^{n}(1-\pi_i)\Phi\left(\frac{\varepsilon_i}{\sigma} \le \frac{X_i\beta - \ln\left(\frac{A_i - t}{t}\right)}{\sigma}\right) + \pi_i\left(1 - \Phi\left(\frac{\varepsilon_i}{\sigma} \le \frac{X_i\beta - \ln\left(\frac{A_i - t}{t}\right)}{\sigma}\right)\right)$$

where $\pi_i = 1$, if *i*th respondent says yes to the question and 0 otherwise. In this case the parameters to be estimated by an ordinary probit are $\frac{\beta}{\sigma}$, linked to X regressor, and $\frac{1}{\sigma}$ associated to regressor $-\ln\left(\frac{A_i - t}{t}\right)$. One can recover β coefficients, present in the willingness to pay function WTP*, through a simple division of these estimated coefficients. The interest in recovering β coefficients lies in the fact that the median and marginal effects appear to be a function of those coefficients.

7.2.3 Non-Parametric Approximations

To avoid the potential problems of parametric measures, and instead of bounding the WTP to zero, or to a upper limit or both, Haab and McConell (1995,1997) propose non-parametric alternatives to obtain lower limits of the WTP mean and median. The two measures that these authors suggest are Turnbull and Kriström.

7.2.3.1 Turnbull Non-Parametric Measure

According to Vaughan, et al (1999) if the bid is b_j, j=0,1,...M+1, $b_j>b_k$ for j>k, $b_0=0$ and the probability of the respondent interviewed WTP lies between the interval b_{j-1} and b_j, then:

$$(7.10)\quad p_j = P(b_{j-1} < WTP \le b_j) \text{ for j} = 1,\ldots,\text{M}+1$$

Also, if the cumulative distribution function is:

$$(7.11)\ F_j = P(WTP \le b_j) \text{ for j} = 1,...\text{M}+1, \text{ where } \text{F}_{\text{M+1}} = 1$$

Then:

$$(7.12)\ \text{p}_j = \text{F}_j - \text{F}_{j-1}$$

and F_0 is equivalent to 0.

Therefore, all the probabilities can be estimated in the following way:

i. Let N_j and Y_j represent the number of "no" and "yes" responses respectively, registered in every *bid*[112] interval.

ii. If $[N_j/(N_j+Y_j)]>[N_{j-1}/(N_{j-1}+Y_{j-1})]$ for all j is between 1 and M, then: $p_j=[N_j/(N_j+Y_j)]-[N_{j-1}/N_{j-1}+Y_{j-1})]$

iii. The probability $N_j/(N_j+Y_j)$ represents the proportion of people interviewed that said "no" to the *bid* b_j, so they are a natural estimator of F_j.

iv. Then, the estimator of p_j would be:

$$(7.13) \qquad p_j = F_j - F_{j-1}, \text{ where, } F_j = \frac{N_j}{N_j + Y_j}$$

v. And the expected WTP would be:

$$(7.14) \qquad E(WTP) = \int_0^\infty WTPdF(WTP) = \sum_{j=1}^{M+1} \int_{b_{j-1}}^{b_j} WTPdF(WTP)$$

Replacing the WTP by the lower limit of each interval, an estimation of the lower limit (LB) of the WTP is obtained:[113]

$$(7.15)\ E(LB_{WTP}) = 0 \cdot P(0 \le w < b_1) + b_1 P(b_1 \le w < b_2) + ... + b_m P(b_m \le w < b_{m+1}) = \sum_{j=1}^{M+1} b_{j-1} p_j$$

Where $\text{p}_{\text{M+1}} = 1 - F_M$

---

[112] To proceed with the calculus the observations were grouped in tariff ranges, therefore the percentage of negative responses will increase according to higher ranges.

[113] See Vaughan et al (1999) for a more complete explanation.

vi. Finally, the lower limit estimation variance would be:

$$(7.16)\ V(\sum_{j=1}^{M+1} b_{j-1} p_j) = \sum_{j=1}^{M+1} b_{j-1}^2 (V(F_j) + V(F_{j-1}) - 2\sum_{j=1}^{M+1} b_{j-1} \sqrt{V(F_j)}\sqrt{V(F_{j-1})}$$

$$\text{Where, } V(F_j) = \frac{F_j(1-F_j)}{(N_j + Y_j)}$$

It is important to mention that the b_M used in the calculation is the maximum offered bid and is the lower limit of the last interval that goes from b_M to infinity, and this is what determines the condition of estimation.

7.2.3.2 Kriström Non-Parametric Measure

Kriström's (1990) non-parametric method consists of grouping the frequency of the "yes" responses to the bid range in a monotonically decreasing order with increasing bid ranges. Next, the points are connected by linear interpolation. To obtain the mean of the WTP the integral below the cumulative density function is approximated as shown in the following equation:

$$(7.17)\ E(x) \text{ in the interval } x_2 - x_1 = \int_{x_1}^{x_2} xf(x)dx = x[F(x_2) - F(x_1)] \text{ for } (x_1 \le x \le x_2)$$

where x_1 and x_2 are the lower and upper limits of bid x, respectively, and f(x) and F(x) are the probability density function and cumulative distribution function, respectively. By this method, the mean willingness to pay is the sum of all the submeans. Using the lower limit of each interval for every bid x_i and applying the above equation for each interval, the mean willingness to pay is obtained according to the following expression:

$$(7.18)\ E(x) \approx x_1[F(x_2) - F(x_1)] + x_2[F(x_3) - F(x_2)] + x_3[F(x_4) - F(x_3)] + x_{n-1}[F(x_n) - F(x_{n-1})]$$

Note that in the case where a survey does not reveal the bid that drives acceptance to zero, the researcher needs to specify it. So, unlike Turnbull's non-parametric measure, Kriström's mean depends in part on this value and hence adds some arbitrariness to the estimation.

7.3 Data and Summary Statistics

The present study uses data from two household surveys conducted in the six different rural regions in Bangladesh and the south rural region of Peru described in Appendix-I. For both countries, the survey followed a two stage random sampling procedure and focused on the respondent's willingness to pay for access to a public telephone service. Both surveys had four different bids for each type of call and used a dichotomous choice format. The survey divided all the respondents into four different groups, and each respondent was randomly assigned to encounter only one of the four bids. In addition to the households' willingness to pay questions, the surveys also contained questions on the households' demographic and housing characteristics, employment, income and expenditure, participation in social programs, and availability of infrastructure, among other characteristics. Both surveys included settlements with and without a public telephone where a lack of a public telephone service was primarily due to a supply constraint instead of one of demand.[114]

Several researchers have recently examined the issue of survey related bias in the contingent valuation willingness to pay setting.[115] Two surveys on which the present study relies do not suffer from sample non-response bias. However, it is necessary to distinguish between unit non-response and item non-response. When a selected respondent does not respond to any of the survey questions, thereby exhibiting unit non-response, and when a respondent does not respond to some of the items of the survey, thereby exhibiting item non-response, estimated WTP suffers from non-response bias. In the present case, there is no unit non-response either in the case of Bangladesh or in the case of Peru, and the extent of item non-response is limited only to 3.2 percent in the case of Peru and a mere 1.4 percent in the case of Bangladesh.

Table-7.3a: Summary Statistics

Covariates	Bangladesh		Peru	
	Mean	Std Dev	Mean	Std Dev
Tariff (in US$)				
3 minutes local call	0.178	0.113	0.370	0.191
1 minute NLD call	0.405	0.265	0.555	0.253
1 minute international call	1.128	0.506		
Total yearly expenditure of HH (in US$)	1455.291	1753.401	2415.297	2093.312
Time required to reach to the phone (in hrs)	0.612	0.585	0.305	0.8936
Distance to the public telephone (in km)	4.887	5.391	9.412	38.397

Note: For Bangladesh figures are for 2001, and for Peru figures are for 1999. The exchange rate for Bangladesh was 1US$=Tk.56. For Peru, the exchange rate was 1US$=3.38S/. Exchange rate source for Peru: WDI 2001.

Table-7.3a presents summary statistics of the tariff, household expenditure, time required to reach the nearest public telephone and distance to the nearest public telephone. Summary statistics for the full set of variables that are included in the estimation are provided in Appendix-IV in table A.7.3a and table-A.7.3b. Among the covariates included in the parametric estimation, tariff and household expenditure are measured in local currency, time required to

[114] Many of the settlement inhabitants without access to a public telephone service travel many kilometres to arrive at a settlement that has the service.

[115] See Messonnier et al (2000) for a discussion on this issue. See also the references therein.

reach the nearest public telephone in hours and distance to the nearest telephone in kilometres.

Other covariates that have been included in the case of Bangladesh are household characteristics such as gender, age, age square, occupation and education of the head of the household, household's migration characteristics, and alternative communication means. In the case of Peru, additional covariates are advantages and disadvantages of public telephones, the cost of alternatives to public telephones, settlement dummies such as educational and health personnel, duration of calls and a public telephone dummy. Summary statistics of these covariates are provided in table-A.7.3a and table-A.7.3b

Turning to the response of surveyed households in the case of Bangladesh, the survey focused on the respondent's willingness to pay for three different types of telephone call: i) local call, ii) national long distance call, and iii) international call. Table-7.3b to table-7.3d show the responses according to the call type. The settlements are divided into with and without telephone access where a settlement with telephone access is defined as the distance of the nearest telephone within a range of 2 kilometres of the settlement.

In the case of Peru, the survey focused on the respondent's willingness to pay for two different types of telephone call: i) local call and ii) national long distance call. Table-7.3e and table-7.3f show the responses according to the call type. The reason for exclusion of international calls in the case of Peru is already described. As usual, the households are divided based on settlements with and without telephone access where a settlement with telephone access is defined as a settlement having a public telephone and not otherwise.

Table-7.3b: Bid Groups and WTP for a Three-Minutes Local Call in Bangladesh

Bid Groups (in local currency, Tk.)	TOTAL	Settlement with Telephone access (%)	Settlement without Telephone access (%)
Bid 1 (Tk. 1.7)			
Do not want to pay	2.74	0.00	5.13
Prefer to pay	97.26	100.00	94.87
Bid 2 (Tk. 7)			
Do not want to pay	29.58	43.24	14.71
Prefer to pay	70.42	56.76	85.29
Bid 3 (Tk. 13)			
Do not want to pay	60.00	66.67	55.56
Prefer to pay	40.00	33.33	44.44
Bid 4 (Tk. 18)			
Do not want to pay	74.36	84.44	60.61
Prefer to pay	25.64	15.56	39.39

Exchange rate, as of January-February 2001, was US$1=Tk.56.00; the price of a three minute local call varies from US$0.03 to US$0.32.

Table-7.3c: Bid Groups and WTP for a One-Minute NLD Call in Bangladesh

Bid Groups (in local currency, Tk.)	TOTAL	Settlement with Telephone access (%)	Settlement without Telephone access (%)
Bid 1 (Tk. 3.4)			
Do not want to pay	4.11	5.88	2.56
Prefer to pay	95.89	94.12	97.44
Bid 2 (Tk. 16)			
Do not want to pay	52.86	59.46	45.45
Prefer to pay	47.14	40.54	54.55
Bid 3 (Tk. 29)			
Do not want to pay	85.00	100.00	75.00
Prefer to pay	15.00	0.00	25.00
Bid 4 (Tk. 42)			
Do not want to pay	88.31	100.00	71.88
Prefer to pay	11.69	0.00	28.13

Table-7.3d: Bid Groups and WTP for a One-Minute International Call in Bangladesh

Bid Groups (in local currency, Tk.)	TOTAL	Settlement with Telephone access (%)	Settlement without Telephone access (%)
Bid 1 (Tk. 26)			
Do not want to pay	16.67	23.53	10.53
Prefer to pay	83.33	76.47	89.47
Bid 2 (Tk. 51)			
Do not want to pay	41.43	45.95	36.36
Prefer to pay	58.57	54.05	63.64
Bid 3 (Tk. 75)			
Do not want to pay	68.33	83.33	58.33
Prefer to pay	31.67	16.67	41.67
Bid 4 (Tk. 100)			
Do not want to pay	67.95	75.56	57.58
Prefer to pay	32.05	24.44	42.42

Exchange rate, as of January-February 2001, was US$1=Tk.56.00; the price of a one minute international call varies from US$0.46 to US$1.78.

For local calls, at the lowest level of bid, around 82 percent of the respondents in Bangladesh, and around 89 percent of the households in Peru are willing to accept the bid. As the bid goes up, the percentage of households willing to accept goes down. For national long distance calls, around 66 percent of the households in Bangladesh and around 74 percent of the households in Peru are willing to accept the bid and, similar to local calls, as the bid increases, willingness to accept decreases. A similar trend is observed in the case of international calls in Bangladesh. For all three types of call and for both countries, high bid prices drive the WTP down; as the bid goes up, the percentage of household's willing to accept the bid goes down. However the present surveys neither reveal the lowest bid at which all households are willing to accept, nor the highest bid which drives the acceptance to zero. It is interesting to note that, for both local and long distance national calls both in Bangladesh and Peru, the percentage of households that accept bids at all levels of the bids is higher from the settlements without telephones than from the settlements with telephones.

Table-7.3e: Bid Groups and WTP for a Three-Minutes Local Telephone Call in Peru

Bid Groups (in local currency, Tk.)	TOTAL	Settlement with Telephone access (%)	Settlement without Telephone access (%)
Bid 1: S/. 0.50			
Do not want to pay	12.9	13.8	10.9
Prefer to pay	87.1	86.2	89.1
Bid 2: S/. 1.0			
Do not want to pay	51.7	53.2	48.5
Prefer to pay	48.3	46.8	51.5
Bid 3: S/. 1.50			
Do not want to pay	76.5	81.8	70.7
Prefer to pay	23.5	18.2	29.3
Bid 4: S/. 2.0			
Do not want to pay	84.5	85.4	83.5
Prefer to pay	15.5	14.6	16.5

Exchange rate in 1999 was US$1=S/. 3.38; the price of a three-minute local call varies from US$0.148 to US$0.59.

Table-7.3f: WTP for a One-Minute National Long Distance Call in Peru

Bid Groups (in local currency, Tk.)	TOTAL	Settlement with Telephone access (%)	Settlement without Telephone access (%)
Bid 1: S/. 1.0			
Do not want to pay	25.7	26.8	23.3
Prefer to pay	74.3	73.2	76.7
Bid 2: S/. 1.5			
Do not want to pay	57.9	58.5	56.7
Prefer to pay	42.1	41.5	43.3
Bid 3: S/. 2.0			
Do not want to pay	80.7	85.9	75.0
Prefer to pay	19.3	14.1	25.0
Bid 4: S/. 3.0			
Do not want to pay	86.4	88.3	84.1
Prefer to pay	13.6	11.7	15.9

Exchange rate in 1999 was US$1=S/. 3.38 (year-average); the price of a one-minute national long distance call varies from US$0.296 to US$0.888.

In addition, table-A.7.3c to table-A.7.3e in Appendix-IV provide the surveyed households willingness that divide the respondents further on a poor/ not-poor basis. Here the poor/ not-poor classification is based on the respective country's national poverty line. Controlling for settlements with and without a telephone, the poor/ not-poor classification does not show any additional price sensitivity of the poor compared to the not-poor. Similar to the total sample, the majority of poor households are willing to accept a low bid and, as the bid goes up, the acceptance goes down.

7.4 Estimations and Results

At this stage, the study proceeds in two steps: first, it estimates the parameters; second, it calculates mean willingness to pay using the estimated parameters. For the sake of tractability, they are discussed separately.

7.4.1 Parameter Estimation

The study has estimated parameters for local calls and long distance national calls for Peru and Bangladesh and for international calls for Bangladesh only. For both countries and for all types of call, it has utilized a logit model and a bounded probit model. For the logit model, it has estimated it in two forms, with nominal tariff and in logarithm of tariff. To check the stability of the estimated parameters, it has used three different specifications for each of the models. So, essentially, the current study has estimated 9 regressions for each type of call per country and 45 regressions; table-7.4.1a to table-7.4.1e present the value of the estimated parameters along their standard deviations. The estimated parameters appear to be consistent, and signs of the coefficients are usually conformed to the prior expectations. The full set of estimated coefficients under different specifications is presented in Appendix-IV (tables-A.7.4.1a to table-A.7.4.1m).

Turning to the results, the coefficients of tariff and opportunity cost are significant both for local and long distance national calls for households in Peru and for all three types of call for households in Bangladesh. As expected, the higher the tariff, the less willing the household is to pay for it, implying that rural households are tariff-sensitive in their willingness to pay for access to a public telephone. For opportunity cost that accounts for travelling time and cost to reach the nearest publicly accessible telephone, which is the non-tariff access cost, and which could be substantial in a rural context. As the estimated coefficient of the opportunity cost is significant for all types of call, it appears that households' private time-price cost plays an important role in their willingness to pay for access to a public telephone. As the coefficients suggest, for Peru, the higher the time required to reach the nearest telephone, the higher the households' willingness to pay. This can also be observed in the case of households in Bangladesh. The coefficient of alternative means of communication to the telephone is also significant for local and national long distance calls both for Peru and Bangladesh; the higher the cost of the alternative, the higher the willingness to pay for a telephone. In addition, while households' members' migration to a city does not have any significant impact on WTP for local and LDN calls, households' members overseas migration is significant for international calls for Bangladesh.

There are two noticeable differences between the two countries. First, for local calls in Peru, income is a significant determinant of the households' willingness to pay while it is not statistically significant for the households' willingness to pay in Bangladesh. For Peru, it shows that the local telephone call is a normal consumption good as the household's willingness to pay for a telephone increases with increased income. For Bangladesh, one possible explanation of this phenomenon is the lesser availability of alternative means of communication and the relative high cost of alternatives which make the local call price inelastic. Second, for Bangladesh, the coefficient of education is statistically significant for all three types of call while it is not significant for Peru (it has been dropped from Peru's regression as a result). However, note that in the case of Bangladesh, it is only the literate-illiterate category that is significant, and the marginal contribution of an additional year of schooling is not significant. This implies that a minimum level of schooling is required to demand a telephone, and that once the threshold is passed, there is no significant difference between differences in years of schooling.

Table-7.4.1a: Parameter Estimates for Local Call: Peru
Dependent Variable: Accept to pay bj soles (Standard errors are in the parentheses)

Explanatory Variables	Logit Model	Bounded Probit
Tariff[a)]	-2.8725 ***	-1.80866***
	(0.2015	(0.1088)
Logarithm of the expenditure	0.2808 **	0.17203**
	(0.1411	(0.0814)
Populated rural centre with public phone	-0.3998 **	-0.23986**
	(0.2102	(0.1224)
Time required to arrive at the public telephone (hours)	0.4675 ***	0.26912***
	(0.1027	(0.0578)
Benefits of using telephone		
Fast communication (yes=1, no=0)	0.5967 ***	0.36376***
	(0.2035	(0.1202)
It avoids trips (yes=1, no=0)	0.9683 ***	0.55452***
	(0.2363	(0.1371)
It avoids isolation (yes=1, no=0)	0.1572	0.10651
	(0.2790	(0.1613)
It saves money (yes=1, no=0)	0.2027	0.15309
	(0.3232	(0.1832)
It saves time (yes=1, no=0)	-0.1903	-0.08123
	(0.2458	(0.1460)
Useful in emergency (yes=1, no=0)	0.4710 **	0.27636**
	(0.2304	(0.1293)
Disadvantages in the use of public telephone		
Excessive distance to the public telephone (yes=1, no=0)	-0.5247 **	-0.30923**
	(0.2502	(0.1461)
Costly (yes=1, no=0)	0.0048	0.00277
	(0.2389	(0.1377)
Bad communication (yes=1, no=0)	-0.3505	-0.19373
	(0.3430	(0.1966)
I don't find the person (yes=1, no=0)	0.7363 **	0.36852*
	(0.3561	(0.2050)

Timetable restriction (yes=1, no=0)	-0.2975	-0.14855
	(0.2254	(0.1314)
No disadvantage (yes=1, no=0)	1.0307 *	0.65028*
	(0.6055	(0.3887)
Logarithm of transfers to relative (proxy of migration)	0.0361	0.02045
	(0.0442	(0.0256)
Cost of alternative communication means:		
Need to travel	0.0360 *	0.01817
	(0.0203	(0.0133)
Letter shipment through relatives	-0.0008	-0.00030
	(0.0035	(0.0024)
Shipment charges	0.0020	0.00119
	(0.0131	(0.0074)
Communication by radio	-0.0519	-0.02995
	(0.0402	(0.0249)
Need to travel to the next populated centre	0.0077 *	0.00398
	(0.0046	(0.0031)
Shipment of mail	0.0744	0.04941
	(0.0805	(0.0411)
Duration of the calls	-0.0857	-0.03977
	(0.0796	(0.0468)
Educational personnel in the CCPP	1.1788 ***	0.67465***
	(0.4100	(0.2112)
Health personnel	-0.9885	-0.55601
	(0.6379	(0.3955)
Constant	1.0073	-9.65643***
	(0.8785	(0.7215)
Log likelihood	-456.0	-446.17
Number of observations	97	971
Sigma		-0.553
Wald chi2	233.5	314.28
Probability > chi2	0.00	0.000

[a] For bounded probit, tariff is the logarithm of the upperbound tariff; ***, **, * significant at 1%, 5% and 10%, respectively.

Table-7.4.1b: Parameter Estimates for Local Call: Bangladesh

Dependent Variable: Accept to pay bj Tk. (Standard errors are in the parentheses)

Explanatory Variables	Logit Model	Bounded Probit
Tariff[a]	-0.24200***	-1.16850***
	(0.0291)	(0.1455)
Logarithm of HH expenditure	-0.03430	-0.00590
	(0.20560)	(0.1223)
Opportunity cost of tel use (travel time and travel exp for tel)	0.42320***	0.27740***
	(0.1493)	(0.0888)
Distance of post office (alternative communication means)	-0.14370*	-0.08860*
	(0.08720)	(0.0516)
Family members migrated to city (yes=1)	0.23410	0.13130
	(0.26570)	(0.1466)
Family members migrated abroad (yes=1)	0.15340	0.07630

	(0.13100)	(0.0741)
Household characteristics		
Gender of the head of the household (male=1)	0.14880	0.0978
	(0.66470)	(0.3887)
Age of the head of the household	0.06270	0.05480
	(0.07990)	(0.0478)
Age square of the head of the household	-0.00070	-0.00050
	(0.00080)	(0.0005)
Occupation of the head of the household	0.25210	0.08500
	(0.32720)	(0.1929)
Education of the head of the household	1.07040	0.65040
	(0.4556)**	(0.2644)**
Constant	0.05000	-6.91410
	(2.95050)	(1.7515)***
Log likelihood	-130.62	-127.172
Number of observations	282	282
Sigma		-0.856
Wald chi2	121.48	128.38
Probability>chi2	0.000	0.000

[a)] For bounded probit, tariff is the logarithm of the upperbound tariff; ***, **, * significant at 1%, 5% and 10%, respectively.

Table-7.4.1c: Parameter Estimates for Long Distance National Call: Peru
Dependent Variable: Accept to pay bj soles (Standard errors are in the parentheses)

Explanatory Variables	Logit Model	Bounded Probit
Tariff[a)]	-2.35914***	-2.17996***
	(0.1819)	(0.1480)
Logarithm of the expenditure	0.25008**	0.15139**
	(0.1343)	(0.0792)
Populated rural centre with public phone	-0.62334***	-0.37539***
	(0.2005)	(0.1168)
Time required to arrive at the public telephone (hours)	0.20301**	0.12055**
	(0.1026)	(0.0576)
Benefits of using telephone		
Fast communication (yes=1, no=0)	0.06729	0.03821
	(0.1907)	(0.1120)
It avoids trips (yes=1, no=0)	0.67554***	0.39674***
	(0.2156)	(0.1264)
It avoids isolation (yes=1, no=0)	0.50744**	0.29267**
	(0.2461)	(0.1458)
It saves money (yes=1, no=0)	0.15526	0.09725
	(0.2998)	(0.1745)
It saves time (yes=1, no=0)	-0.27863	-0.18026
	(0.2302)	(0.1368)
Useful in emergency (yes=1, no=0)	0.09451	0.05710
	(0.1973)	(0.1155)
Disadvantages in the use of public telephone		
Excessive distance to the public telephone (yes=1, no=0)	-0.59135**	-0.35077**
	(0.2563)	(0.1476)

Costly (yes=1, no=0)	0.32261	0.19810
	(0.2378)	(0.1357)
Bad communication (yes=1, no=0)	0.16813	0.10604
	(0.3295)	(0.1926)
I don't find the person (yes=1, no=0)	0.75383**	0.42831**
	(0.3028)	(0.1779)
Timetable restriction (yes=1, no=0)	0.17618	0.11533
	(0.2259)	(0.1324)
No disadvantage (yes=1, no=0)	0.87610	0.46863
	(0.8366)	(0.4826)
Logarithm of transfers to relative (proxy of migration)	0.04173	0.02542
	(0.0403)	(0.0238)
Cost of alternative communication means:		
Need to travel	0.05259**	0.02979**
	(0.0213)	(0.0134)
Letter shipment through relatives	-0.00858	-0.00454
	(0.0077)	(0.0040)
Shipment charges	-0.01071	-0.00714
	(0.0107)	(0.0059)
Communication by radio	-0.03079	-0.01718
	(0.0217)	(0.0122)
Need to travel to the next populated centre	0.00667	0.00373
	(0.0051)	(0.0035)
Shipment of mail	0.09268	0.05925
	(0.0733)	(0.0392)
Duration of the calls	0.03197	0.02397
	(0.0779)	(0.0459)
Educational personnel in the CCPP	0.33650	0.20417
	(0.3729)	(0.2084)
Health personnel	-0.27426	-0.11147
	(0.6157)	(0.3883)
Constant	1.69737**	-10.34388***
	(0.8313)	(0.7945)
Log likelihood	-491.11	-488.25
Number of observations	968	968
Sigma		-0.459
Wald chi2	199.1	256.8
Probability > chi2	0.000	0.000

[a] For bounded probit, tariff is the logarithm of the upperbound tariff; ***, **, * significant at 1%, 5% and 10%, respectively.

Table-7.4.1d: Parameter Estimates for National Long Distance Call: Bangladesh

Dependent Variable: Accept to pay bj Tk. (Standard errors are in the parentheses)

Explanatory Variables	Logit Model	Bounded Probit
Tariff[a]	-0.13680***	-1.33900***
	(0.0159)	(0.1441)
Logarithm of HH expenditure	0.39330*	0.23010*
	(0.23040)	(0.1356)
Opportunity cost of tel use (travel time and travel exp)	0.26720*	0.21440**

	(0.16340)	(0.1003)
Distance of post office (alternative communication means)	0.22210**	0.13610**
	(0.0955)	(0.0553)
Family members migrated to city (yes=1)	0.12100	0.07850
	(0.18730)	(0.1146)
Family members migrated abroad (yes=1)	0.10680	0.06330
	(0.16420)	(0.094)
Household characteristics		
Gender of the head of the household (male=1)	0.62950	0.4098
	(0.74960)	(0.4857)
Age of the head of the household	0.03790	0.06390
	(0.08950)	(0.0556)
Age square of the head of the household	-0.00040	-0.00060
	(0.0009)	(0.0006)
Occupation of the head of the household	-0.17730	-0.18630
	-0.36480	(0.2191)
Education of the head of the household	1.03310**	0.69130**
	(0.5236)	(0.3430)
Constant	-5.23830	-9.94550***
	(3.20550)	(1.9754)
Log likelihood	-109.458	-98.942
Number of observations	280	280
Sigma		-0.747
Wald chi2	164.07	185.11
Probability>chi2	0.000	0.000

[a] For bounded probit, tariff is the logarithm of the upperbound tariff; ***, **, * significant at 1%, 5% and 10%, respectively.

Table-7.4.1e: Parameter Estimates for International Call: Bangladesh

Dependent Variable: Accept to pay bj Tk. (Standard errors are in the parentheses)

Explanatory Variables	Logit Model	Bounded Probit
Tariff[a]	-0.037***	-1.1472***
	(0.0057)	(0.1631)
Logarithm of HH expenditure	0.5703***	0.3199***
	(0.2054)	(0.1160)
Opportunity cost of tel use (travel time and travel exp)	0.3831***	0.2346***
	(0.1422)	(0.0832)
Distance of post office (alternative communication means)	0.1089	0.0632
	(0.08050)	(0.0477)
Family members migrated to city (yes=1)	-0.0068	-0.0004
	(0.17990)	(0.1111)
Family members migrated abroad (yes=1)	0.75***	0.4083***
	(0.2112)	(0.1073)
Household characteristics		
Gender of the head of the household (male=1)	0.1099	0.0313
	(0.64560)	(0.3785)
Age of the head of the household	0.1113	0.0735
	(0.07720)	(0.0467)
Age square of the head of the household	-0.001	-0.0007
	(0.00080)	(0.0005)
Occupation of the head of the household	0.3171	0.1632

	(0.31020)	(0.1843)
Education of the head of the household	0.4914	0.3121
	(0.44050)	(0.2656)
Constant	-8.6947***	-8.1242***
	(2.9379)	(1.6212)
Log likelihood	-143.381	-141.819
Number of observations	280	280
Sigma		-0.872
Wald chi2	101.04	104.17
Probability>chi2	0.000	0.000

[a)] For bounded probit, tariff is the logarithm of the upperbound tariff; ***, **, * significant at 1%, 5% and 10%, respectively.

7.4.2 Willingness to Pay Estimation

Table-7.4.2a to table-7.4.2c report parametric and non-parametric estimates of households' willingness to pay. For the Logit model, estimates are the untruncated mean, the mean truncated at zero but untruncated from above, and the truncated mean confined between zero and the maximum bid. For the bounded Probit model, the tables report the mean and median WTP. The tables also report two nonparametric alternative estimates of mean and median WTP. In addition, the full set of estimated WTP under different specifications of parametric models is presented in Appendix-IV (tables-A.7.4.2a to table-A.7.4.2m). The details of the nonparametric means are provided in tables-A.7.4.2n to A.7.4.2w.

In Peru, rural households' WTP for a three-minutes local call from a public telephone varies from US$0.25 to US$0.35 (0.85 soles to 1.18 soles), and for an LDN call varies from US$0.33 to US$0.45 (1.13 soles to 1.52 soles). In the case of rural households in Bangladesh, households' WTP for a three minute local call from a publicly accessible telephone varies from US$0.10 to US$0.26 (Tk. 5.73 to Tk.14.3), for a one minute LDN telephone call from US$0.23 to US$0.50 (Tk.12.67 to Tk.27.84), and for an international call from US$0.93 to US$1.35 (Tk.51.92 to Tk.75.33).

The estimated total benefits of access to a telephone for rural households depend largely on which method one employs in the estimation process. Given the difference in benefits figures from the estimation of WTP, which estimated WTP figure represents the actual project benefits of a rural telephony project in Bangladesh or in Peru? Habb and McConnell (1997) argue for the median WTP, as it is less sensitive to distributional misspecifications and estimation methods. However, all the benefit measures can be considered as equally valid given that they at least give a range over which a scenario of potential benefits can be built.

The variation of benefits and their magnitudes show that it is not possible to find a single number that correctly represents households' willingness to pay for access to a public telephone. However, the net benefits of access to a telephone remain positive irrespective of the estimation method. The estimates of both parametric and non-parametric measures of WTP for all types of call for both Bangladesh and Peru suggest that rural households are willing to pay at least as much as the prevailing tariff rate to get access to the public telephone. In fact, households' WTP for local and LDN calls in Peru, and local, LDN and international calls in Bangladesh is higher than the prevailing tariff rate. Though the different estimates of WTP that have been employed here vary from each other, each of the WTP estimates is higher than the existing tariff rate, making households' lowest WTP higher than the level of existing tariff.

Table-7.4.2a: Households' WTP for Local Calls (in US$)

		Peru	Bangladesh
Parametric Estimation-Logit			
Mean WTP	-infinity<WTP<+infinity	0.337	0.219
		(0.009)	(0.011)
Truncated Mean WTP	0<WTP<+infinity	0.341	0.223
		(0.009)	(0.011)
Truncated Mean WTP	0<WTP< Bmax	0.252	0.178
		(0.004)	(0.006)
Parametric Estimation-Probit			
Median WTP		0.300	0.189
		(0.008)	(0.014)
Mean WTP		0.350	0.255
		(0.022)	(0.025)
Non-parametric-Turnbull			
Mean WTP		0.264	0.162
		(0.002)	(0.036)
Median WTP			0.102
Non-parametric-Kriström			
Mean WTP		0.324	0.209
		(0.015)	(0.035)
Median			0.160

Numbers in the parentheses are the standard errors. Standard errors are computed through bootstrap procedure based on 100 replications for each sample.

Table-7.4.2b: Households' WTP for LDN Calls (in US$)

		Peru	Bangladesh
Parametric Estimation-Logit			
Mean WTP	-infinity<WTP<+infinity	0.428	0.331
		(0.011)	(0.025)
Truncated Mean WTP	0<WTP<+infinity	0.432	0.341
		(0.010)	(0.023)
Truncated Mean WTP	0<WTP< Bmax	0.340	0.327
		(0.007)	(0.021)
Parametric Estimation-Probit			
Median WTP		0.409	0.255

	(0.009)	(0.018)
Mean WTP	0.451	0.320
	(0.016)	(0.022)
Non-parametric-Turnbull		
Mean WTP	0.333	0.226
	(0.010)	(0.069)
Median WTP		0.265
Non-parametric-Kriström		
Mean WTP	0.416	0.337
	(0.013)	(0.054)
Median		0.497

Numbers in the parentheses are the standard errors. Standard errors are computed through bootstrap procedure based on 100 replications for each sample.

Table-7.4.2.3: Households' WTP for International Calls (in US$)

		Bangladesh
Parametric Estimation-Logit		
Mean WTP	-infinity<WTP<+infinity	1.210
		(0.108)
Truncated Mean WTP	0<WTP<+infinity	1.248
		(0.104)
Truncated Mean WTP	0<WTP< Bmax	0.957
		(0.034)
Parametric Estimation-Probit		
Median WTP		1.099
		(0.106)
Mean WTP		1.345
		(0.148)
Non-parametric-Turnbull		
Mean WTP		0.927
		(0.239)
Median WTP		1.065
Non-parametric-Kriström		
Mean WTP		1.149
		(0.261)
Median		1.203

Numbers in the parentheses are the standard errors. Standard errors are computed through bootstrap procedure based on 100 replications for each sample.

7.5 Conclusions

This chapter has measured the welfare impact of the availability of a telephone on rural households through the measurement of Hicksian equivalent variation. It has used both parametric and non-parametric methods in estimating rural households' willingness to pay for access to a public telephone in a simple referendum setting. In contingent valuation, the use of the referendum method has become the presumptive approach. One major reason for this wide acceptance is that the referendum model appears to be incentive compatible and thus does not suffer from strategic behaviour in revealing the truth, which is a major concern with contingent valuation.

Findings of both parametric and non-parametric estimation suggest that such projects are welfare enhancing. However, despite the fact that the exact welfare gain of such projects varies depending on the methods employed to estimate the benefit, rural households' welfare gain remains positive across all the estimation methods.

This evidence coupled with the evidences provided in the previous chapter confirms that the two alternative institutions described in the first part of study have a direct welfare impact on the rural households and the impact is positive. For the indirect impact of telephones on rural households, the next two chapters provide necessary evidences.

8 MEASURING THE INDIRECT IMPACTS: ACCESS TO INFORMATION AND FACTOR MARKET PARTICIPATION

8.1 Introduction

This chapter and the subsequent chapter measure the indirect impacts of the use of telephone on rural households. While this chapter examines the impact of access to information on factor market participation, the subsequent chapter examines the impact of access to information on product market participation. While the critical role of information as a cause of market imperfection in developing countries has become axiomatic, the theory is backed by little empirical evidence on the magnitude of the problem. The present study asks whether access to information through access to telecommunications brings any discrete change in a farm household's land and labour market participation decisions. More specifically it asks two questions. First, does the use of a telephone have any impact on a farm household's factor market participation decision? Second, correcting for market participation, does the use of a telephone have any impact on the type of factor adjustment decision?

In most developing countries, agriculture remains a major contributor to GDP in general, and it remains the prime source of livelihood of rural people in particular.[116] Despite different policy interventions, the sector has remained semicommercialized, dominated by self-sufficient family farms.[117] Many markets, particularly future markets for risks and capital are either absent or extremely limited, small and illiquid. Even when a market exists, its use is not universal. An analysis of the participation decision is important as understanding of the household's participation decision increases the available set of policy instruments that can influence the household's behaviour. As market participation is assumed to increase efficiency in resource allocation, understanding of the factors that influence the participation decision can help to design policy aimed at improving household welfare.

Land and labour transactions usually increase resource allocation efficiency as agents with high expected marginal productivity of land acquire land and vice versa, and agents with high expected productivity of labour hire labour and vice versa. In this way, a downward (upward) adjustment of labour or an upward (downward) adjustment of land are the alternative means of exploiting resource endowments in agrarian economies. However, though beneficial, not all households participate in such factor adjustments. Among many reasons that might cause selective market participation, market imperfection due to

---

[116] In the case of least developed countries, agriculture value added as a percentage of GDP was 37% in 1997. Least developed countries (UN classification) include 47 economies. Source, World Development Indicators, World Bank.

[117] As of 1981, the percentage of owner-cultivated farms was 79% of total farms. See, United Nations Food and Agriculture Organization (1981) cited in Otsuka et al (1992), p. 1971.

information imperfection is one of the prominent ones. The study thinks transaction cost as a cause of selective factor market participation and considers information cost as a part of transaction cost.

Transaction cost is a widely used approach to explain the observed market failures and self-sufficiency in agriculture in developing countries.[118] One fundamental source of transaction cost is information cost. When information is not costless, this has important implications in present and future contracts and transactions. Transactions and contracts, which could be feasible in the presence of perfect information, may not occur.[119] The existence of transaction cost generates a wedge between a household's buying and selling price,[120] i.e., once adjusted for transaction costs, the rental price of land and labour can be different depending on whether the household is on the demand or on the supply side of the market. As a result, transaction costs reduce the market size and, in extreme cases, when transaction costs are very high, the market may fail.

The strand of economic literature that recognizes the possibility of market imperfections usually identifies asymmetries of information and problems of moral hazard as the primary reasons for market failure in agriculture in developing countries.[121] In many instances, agrarian institutions that govern the transactions of land and labour are endogenous as they adapt to reflect information costs.[122] For example, in the absence of perfect information, sharecropping is a rational response to overcome incentive problems caused by informational imperfections where an agent's effort cannot be directly monitored by the principal.[123] However, if land and labour markets do not function properly only because of costly and imperfect information, then an increase in information availability may change the functioning of markets.

As mentioned in Chapter 1, the present study has taken the use of the telephone as a proxy for access to information, and has assumed a positive relationship between availability of information and market participation. Recent expansion of telecommunications services to rural areas in Bangladesh has increased the access of rural households to information. If availability of information increases market participation, then the use of the telephone may increase market participation because the use of the telephone provides increased access

---

[118] See Key et al (2000) for a theoretical analysis and empirical evidence.

[119] See Akerlof (1970).

[120] See Hirshleifer (1984), pp. 421-23.

[121] See for example, Binswanger and Rosenzweig (1984), Stiglitz (1988), Bardhan (1989), and Otsuka et al (1992).

[122] See Stiglitz (1988), p. 100.

[123] See Stiglitz (1974) that assumes that a landlord cannot monitor the effort of labour. In addition, this principal-agent paradigm comprises a large body of literature. For a recent application in an agrarian context, see Otsuka et al (1992).

to information and hence its use reduces information imperfection. Access to telecommunications has potentials to overcome limitations associated with information imperfection, and to reduce transaction costs. 'Two key determinants of market emergence are the costs associated with acquiring information, and the cost of negotiating transactions';[124] and the spread of telecommunications is expected to reduce both acquiring and negotiation costs. With this background, the present study examines the impact of transaction costs on a farm household's factor market participation where search and information cost is a part of transaction costs.

Though selective, adjustments of land and labour margins of agricultural households in Bangladesh are not uncommon. In the case of land, observations show that there are both short-term adjustments through the transfers of user-rights, such as renting out or sharecropping, and long-term adjustments through the transfer of ownership such as selling. Similarly, adjustments of labour margins in the form of temporary, seasonal and permanent contracts can also be found. The observed market participation behaviour of households complies with the view that market participation is household specific, as argued by de Janvry et al (1991), and participation is not factor specific. Markets for factors exist, but not all farm households utilize them.

Table-8.1 shows the observed factor adjustment of agricultural households in Bangladesh and table-8.2 reports the market participation behaviour of households for telephone users and nonusers. The figures are generated from a primary survey conducted in Bangladesh in the year 2001. The survey methods and coverage will be discussed in section-8.4. Among the surveyed households, more than 62 percent of them adjust land through the market, and more than 64 percent of them adjust labour through the market.

Table-8.1: Observed Factor Adjustment of Agricultural Households in Bangladesh

Factor Market Participation	Land		Labour	
Remain in autarky	107	(37.68)	102	(35.92)
Participate in factor market	177	(62.32)	182	(64.08)
Hired-in/Leased-in	68	(23.94)	131	(46.12)
Hired-out/Leased-out	56	(19.72)	35	(12.32)
Both in and out	53	(18.66)	16	(5.63)
Total	284	(100.00)	284	(100.00)

Numbers in the parentheses are the percentages; source: primary survey.

Turning to the effect of telephone use, one can find that the raw difference in factor market participation rate between households that use a telephone and households that do not use a telephone is about 14 percent (table-8.2). The last row of table-8.2 presents the mean of participation and for users and nonusers

[124] See Leff (1984).

of telephones and their difference where the difference is significant at the level of one percent. Examining the patterns of market participation rates for households with respect to telephone use, it appears that the use of a telephone has had an important effect on market participation behaviour.

Table-8.2: Use of Telephone and Factor Market Participation

Factor Market Participation	Telephone Use Status				Total (%)	
	Yes (%)		No (%)			
No	5	(5.81)	39	(19.70)	44	(15.49)
Yes	81	(94.19)	159	(80.30)	240	(84.51)
Total	86	(100.00)	198	(100.00)	284	(100.00)
Mean of participation	0.942	(0.235)	0.803	(0.399)	0.139**	(0.046)

Numbers in the parentheses are the percentages and standard deviations/errors; **the mean difference is significant at 1% level. Source: primary survey.

This chapter contributes to the existing literature in two ways: first, it estimates empirically the impact of access to information in the form of the use of a telephone on a farm household's market participation behaviour. In doing so, it controls for the possible self-selection bias by jointly estimating both the propensity of a household to participate in the market and the match of households to telephone use. Second, the chapter estimates empirically the impact of access to information on specific factor adjustment. At this stage, it estimates a farm household's probability to participate in a specific market correcting for participation. The remainder of the chapter proceeds as follows: section 8.2 provides a descriptive review of land and labour adjustment patterns of farm households in Bangladesh and relates the observed pattern with the telephone use; section 8.3 constructs a framework that provides a theoretical basis for market participation decisions; section 8.4 describes the data collection methods and descriptive statistics; section 8.5 provides the empirical estimation; and section 8.6 concludes.

8.2 Land and Labour Adjustments: An Overview

This section starts with examining the relationship between a household's factor endowments and factor market participation. It also examines the simultaneousness of adjustments of land and labour. Later, it relates factor market participation to the use of the telephone.

Table-8.3 reports the availability of family labour and adjustments of labour margins, and table-8.4 reports the availability of family land and adjustments of land of the surveyed farm households in Bangladesh. It seems that a household's labour adjustments, either hired-in or hired-out, do not depend on the availability of family labour. In contrast to labour, households that lease-out land are mostly big farmers and households that lease-in land are small farmers; while it is only about 10 percent of the small farm households that adjust land downwards, it is about 43 percent of the large farmers that adjust land downwards. This observed relationship between farm size and input adjustment direction is further reinforced by observations on upward adjustment, which are in fact the mirror image of downward adjustment (see table-8.4).

Table-8.3: Availability of Family Labour and Labour Market Participation

Family size	Labour market participation									
	Remain in autarky		Hired-in		Hired-out		Both hired-in and out		Total	
1-4	31	(34.44)	43	(47.78)	11	(12.22)	5	(5.56)	90	(10000)
5-6	51	(38.35)	53	(39.85)	19	(14.29)	10	(7.52)	133	(100.00)
7+	20	(32.79)	35	(57.38)	5	(8.20)	1	(1.64)	61	(100.00)
Total	102	(35.92)	131	(46.12)	35	(12.32)	16	(5.63)	284	(100.00)

Numbers in the parentheses are the percentages; source: primary survey.

Table-8.4: Availability of Family Land and Land Market Participation

Farm size (in decimals)	Land market participation									
	Remain in autarky		Leased-in		Leased-out		Both leased-in and out		Total	
0-<100	63	(40.10)	48	(30.60)	16	(10.20)	30	(19.10)	157	(100.00)
100-300	31	(33.70)	20	(21.70)	25	(27.20)	16	(17.40)	92	(100.00)
>300	13	(37.10)	0	(0.00)	15	(42.90)	7	(20.00)	35	(100.00)
Total	107	(37.68)	68	(23.94)	56	(19.72)	53	(18.66)	284	(100.00)

Numbers in the parentheses are the percentages; source: primary survey.

Factor adjustment decisions are not independent as a labour adjustment decision is not independent of land adjustment decisions and vice versa. As a result, land market constraints can affect labour market participation and vice versa. For instance, the presence of potential supervision and monitoring constraints on hired labour can limit the upward adjustments of land. Similarly, the presence of potentially binding constraints on off- farm works can limit the downward adjustments of land. In the absence of any constraints, the factor adjustment decision should be independent of its own factor endowments.

Table-8.5 combines both land and labour adjustments. The observed patterns show that about 85 percent of the households adjust at least one of the factor margins, and the remaining 15 percent do not make any factor adjustments. About 20 percent of the households adjust their land margin only, and about 22 percent of the households adjust their labour margin only. About 13 percent of the households adjust both land and labour margins upwards, while none of the households adjust both margins downwards. Note that the observed factor adjustments reported in table-8.5 are for a cropping season for land and for a year for labour. As a result, these adjustments are essentially short term in nature. Households that have not adjusted in the present sample in the short run might already have adjusted in the long run through mechanisms like migration and land selling and have achieved desired factor ratios. To account for such adjustments, migration and land buying/selling will be taken in to consideration.

Table-8.5: Adjustments of both Margins: Cross Tabulations

Land→ **Labour↓**	Remain in Autarky		Lease Land-in		Lease Land-out		Both lease-in and out		Row totals	
In Autarky	44	(15.49)	14	(4.93)	38	(13.38)	6	(2.11)	102	(35.92)
Hire labour-in	44	(15.49)	38	(13.38)	17	(5.99)	32	(11.27)	131	(46.13)
Hire labour-out	15	(5.28)	8	(2.82)	0	(0.00)	12	(4.23)	35	(12.32)
Hire-in and out	4	(1.41)	8	(2.82)	1	(0.35)	3	(1.06)	16	(5.63)
Column totals	107	(37.68)	68	(23.94)	56	(19.72)	53	(18.66)	284	(100.00)

Numbers in the parentheses are the percentages; source: primary survey.

To examine the relationship between adjustments of two margins, the study looks at the conditional probabilities given by table-8.6. The conditional adjustment probabilities show that households unable to adjust one margin are more likely to adjust in the other margin; households that have not adjusted land margins are likely to adjust in the labour margin and vice versa. In addition to this substitutability between adjustments in either of the margins, the probabilities of simultaneous adjustment are also high; given that a household has adjusted its land margin, it is more likely that it will adjust its labour margin as well.

Table-8.6: Land and Labour Adjustments: Conditional Probabilities

Labour, given land:	$(land = 0)$	$(land \neq 0)$
$P(labour = 0 \mid land)$	0.41	0.33
$P(labour \neq 0 \mid land)$	0.59	0.67
Land, given labour:	$(labour = 0)$	$(labour \neq 0)$
$P(land = 0 \mid labour)$	0.43	0.35
$P(land \neq 0 \mid labour)$	0.57	0.65

Table-8.7 shows the probabilities of market participation, given telephone use. The probability that a household uses a telephone and does not participate in the market is very low compared to the probability that a household does not use telephone and does not participate in the market. The probability of participating in the labour market conditional on telephone use is 0.128 while the probability of participating in the land market conditional on telephone use is 0.326. With the exception of labour market participation, it seems from table-8.7 that households that use a telephone have a higher probability to participate in the market than households that do not use a telephone.

Table-8.7: Use of Telephone and Factor Market Participation: Conditional Probabilities

Market participation	Telephone use	
	$(telephone = 0)$	$(telephone = 1)$
$P(mkt = 0 \mid telephone)$	0.197	0.058
$P(labour = 1 \mid telephone)$	0.263	0.128
$P(land = 1 \mid telephone)$	0.152	0.326
$P(land \,\&\, labor = 1 \mid telephone)$	0.389	0.488

Note: 1 indicates that a household participates and/or uses a telephone and 0 indicates the opposite.

The explorative type of analysis of this section suggests that, though selective, farm households in Bangladesh participate in land and labour markets and adjust their factor endowments. However, there exists simultaneity in factor adjustments behaviour among the participating households. In addition, the use of a telephone might have some relationship with the farm household market participating behaviour as households that use a telephone have a higher probability to participate in the market than the households that do not use a telephone.

8.3 Theoretical Framework

Household-farm models usually maintain a common framework where a household maximises its utility function under production, time and budget constraints.[125] Under the presence of transaction costs, henceforth TC, a household's market participation depends on comparison between two utility levels: the utility level that the household can attain when it participates in the factor market net of transaction costs, and the utility level it attains when the household remains in autarky. In the second case, the household does not incur any transaction costs.

Two types of transaction costs are proportional transaction cost (PTC) and fixed transaction cost (FTC). The first one, PTC, depends on the volume of trade, which means TC is incurred in each unit of trade. In comparison with PTC, FTC is specific to the frequency of trade, which means TC is incurred in each trading. In reality, a single trade can be subject to both types of TCs and, despite this distinction, both play the same role: by widening the price gap between buyers and sellers, they (PTC and FTC) reduce exchange. However, the distinction is important in respect of market participation decisions for farm households; while PTC influences the amount to be supplied in the market once a farm household is already in the market, FTC influences whether a farm household participates in the market or not.

Information cost can be viewed as a form of fixed transaction cost. It does not change the unit price that a farm household receives as a supplier or pays as a

[125] See Strauss (1984).

buyer; it is fixed and equally incurred by households that supply land and/or labour and households that demand land and/or labour. As a fixed cost, it delays a household's market participation decision and, in extreme cases, when the problem of information is widespread, a market could be completely missing.

Following Key et al (2000), it is possible to write the household's indirect utility V as a function of market prices, (w,r), the household's income before incurring transaction costs, y, and the household's exogenous utility shifters, Z_h:

$$V = V(w,r,y,Z_h) \qquad (8.1)$$

The utility levels to be compared are when the household adjusts its land and/or labour margins in any direction (buyer or supplier), and when the household remains in autarky. That means

$$V^m = V(w,r,y-tc,Z_h) \qquad (8.2)$$

$$V^a = V(\tilde{w},\tilde{r},\tilde{y},Z_h) \qquad (8.3)$$

Where V^m and V^a are two respective utility levels when the household adjusts its margin or remains in autarky, and $\tilde{w}$, $\tilde{r}$ and $\tilde{y}$ are the autarkic wage, rent, and income respectively. Intuitively, a household will participate in the market when $V^m \geq V^a$, or in other words, $y-tc \geq \tilde{y}$.

8.4 Data Collection Methods and Descriptions

The chapter uses data from the household survey conducted in six different regions of Bangladesh during the months of January and February of the year 2001. The survey is described in the Appendix-I and mentioned in the Chapter 1. In the survey, households were stratified on the basis of the ease of access to telecommunication infrastructure, where access was defined in terms of distance that needs to be travelled to access telephone services. The survey contained questions on the households' production and factor market participation behaviour, and the variables related to access to and use of telecommunications. In addition, the survey also collected respondents' personal and family characteristics. The survey details are described in appendix-1. Table-8.8 provides summary statistics of the variables relevant for the present chapter. For ease of discussion, variables are arranged under three categories: a household's characteristics, a household's factor endowments, and information availability.

In the case of household characteristics, age is measured in years and education is measured in intervals and has been recorded in seven different categories.

Among the head of the households, about 43 percent of them are functionally illiterate and about 7 percent of them have studied at tertiary level. The modal occupation of the head of the households is agriculture (35.92 percent), followed by business/trade (19.01 percent). The total yearly household expenditure is measured in local currency. Table-8.8a and table-8.8b show the descriptive statistics about education and occupation of the head of the households respectively.

Table-8.8: Summary Statistics

Variables	Mean	Std.
1. Household Characteristics		
Gender of the household head (Female=0)	0.944	0.231
Age of the household head	46.567	12.737
Education of the household head (illiterate 0, others 1)	0.838	0.369
Occupation of the household head	0.535	0.500
Log of Yearly household expenditure	10.946	0.787
2. Factor Endowments		
Farm size (owned)	154.581	246.195
Household's labour supply	3.785	1.666
Log of average price of land	10.147	0.692
Log of average wage of labour	4.248	0.183
Household member migrated to city	0.130	0.337
Household member migrated abroad	0.180	0.385
3. Information availability		
Telephone use status (users 1, others 0)	0.303	0.460
Distance of the nearest telephone in KM	4.887	5.391
Telephone users as a percentage of sample size	30.282	19.716
Village type (village with telephone=1)	0.158	0.366

Table-8.8a: Education of the Head of the Household

Level of Education	Frequency	Percent
Illiterate	46	16,20
Can sign only	76	26,76
From Class 1 to 5	48	16,90
From Class 6 to 10	58	20,42
Secondary School Certificate	29	10,21
Higher Secondary Certificate	6	2,11
Graduate	21	7,39
Total	284	100,00

In the case of factor endowments, farm size is the amount of land a household owned; it is measured in decimals. A household's labour supply consists of the number of persons per household between the ages of 14 and 60 years. As both females and males participate in agriculture in one way or another, the present study has not excluded female members of the household in aggregating the household's total labour supply. The average price of land is the average price of farmland and consists of the prices of best and worst land at village level. Similarly, the average wage is the village level agricultural wage per day per person. Both prices are measured in local currency. Among the respondent households, migration to a city or abroad is a visible characteristic. Out of 284 households, 13.03 percent and 17.96 percent of the households have one of or more family members migrated to city or abroad respectively (table-8.8c).

Table-8.8b: Occupation of the Head of the Household

	Frequency	Percent
Taking Care/Managing the household	27	9,51
Agriculture (crop)	102	35,92
Agriculture (non-crop)	2	0,70
Rural Industry	4	1,41
Agriculture Labour	19	6,69
Non-agriculture Labour	10	3,52
Transport	6	2,11
Govt./Private Service	40	14,08
Business/Trade	54	19,01
Others	18	6,34
Total	284	100,00

Table-8.8c: Distribution of Household Members' Migration Status

Family members migrated to city	Frequency	Percent
No	247	86,97
Yes	37	13,03
Total	284	100,00
Family members migrated abroad		
No	233	82,04
Yes	51	17,96
Total	284	100,00

In the case of information availability, telephone use status measures whether any member of the household uses a public or a private telephone. The survey contained questions about present as well as past telephone use status. The distance to the nearest telephone is measured in kilometres, and the telephone user as a percentage of sample size is the number of telephone users in a village as a percentage of the number of respondents from that village. The village type is a dummy variable and is one for a village with a telephone and zero otherwise.

8.5 Empirical Estimation

In the empirical estimation, the present study employs two basic measures of factor market participation. The first one is the factor market participation status that equals 1 if a household has adjusted any of its factor endowments in the last growing season. The second measure is the type of factor adjustment correcting for market participation. At this stage, it considers three possibilities that a household faces: adjust land only, adjust labour only, and adjust both land and labour. In order to be more tractable it discusses them under two different subsections. As usual, it starts each subsection with an empirical specification followed by the estimation and results.

8.5.1 Factor Market Participation Status

For market participation, here, the approach is to augment a standard cross-sectional market participation function to include a dummy indicating whether a household uses a telephone or not. Let T_i represent a dummy variable that equals 1 if the *i*th household uses a telephone, and zero otherwise. Household *i*'s market participation decision is assumed to depend on T_i, and a vector of observed characteristics X_i, and can be described by the latent variable model:

(8.4) $$M_i^* = \alpha' T_i + \beta' X_i + \varepsilon_{1i},$$

where M_i^* is the expected utility level that a household attains from market participation, ε_{1i} is a normally distributed error term with mean zero and unit variance, and α and β are parameters to be estimated.

However, one cannot observe the expected utility level; what one rather observes is whether or not a household participates in factor markets. Assume for each household *i* a dummy variable M_i that expresses the bivariate state of a household's market participation. The conditional probability that for household *i* given T_i and X_i, one observes $M_i = 1$, (participate in factor market) is given by

(8.5) $$P(M_i = 1 | T_i, X_i) = P(\alpha' T_i + \beta' X_i + \varepsilon_{1i} \geq 0) = F(\frac{\alpha' T_i + \beta' X_i}{\sigma_1})$$

The study estimates the probabilities of market participation given T_i and X_i, where T_i and X_i include a household's telephone use status and characteristics, respectively. Table-8.9 reports the normalized probit coefficients estimated with varying sets of covariates. In the first column, telephone use is the only explanatory variable along with a constant term. The raw difference in factor market participation rate between households that use a telephone and households that do not use a telephone is about 14 percent (marginal coefficient of telephone). In column 2, in addition to telephone use, household characteristics have been added as additional covariates. Inclusion of household characteristics reduces the difference between users and non-users to around 11 percent. Even after controlling for additional covariates, the telephone dummy variable continues to have a sizeable impact on a household's factor market participation. Column 3 includes a household's factor endowments. The difference reduces further to 6 percent once controlled for a household's factor endowments in addition to the household characteristics. However, inclusion of these factors does not alter the significance of telephone use.

Table-8.9: Effects of Telephone Use on Factor Market Participation
Dependent variable: Factor market participation status (0,1)
Normalized probit coefficients (marginal effects)

	Model 1	Model 2	Model 3
Telephone use status (yes=1, No=0)	0.1388	0.1071	0.063
	(0.03789)**	(0.0309)**	(0.0258)*
Household Characteristics:			
Gender, household's head (Male=1)		0.1075	0.0904
		(0.1679)	(0.1679)
Age, household's head		0.0105	0.0157
		(0.0065)~	(0.0064)*
Age square		-0.0001	-0.0001
		(0.0001)	(0.0001)*
Education (literate=1)		0.0643	0.0238
		(0.0593)	(0.0348)
Occupation (Agriculture=0)		0.1136	0.0782
		(0.0381)**	(0.0326)**
Ln of household expenditure		0.0093	0.0042
		(0.0213)	(0.0148)
Log of farm size			0.0258
			(0.0098)**
Log of stock of labour			-0.1053
			(0.0392)**
Number of observations	284	284	284
Log likelihood	-117.3185	-78.8387	-65.1347
LR chi2	10.27	87.22	114.63
Probability >chi2	0.0014	0.0000	0.0000
Pseudo R2	0.0419	0.3562	0.4681

Standard errors are in the parentheses. ~Values are probit coefficients, not marginal effects.
**, *: Significant at 1%, and 5% respectively.

Correction for possible omitted variable bias: The estimation results suggest that the use of a telephone has a significant positive impact on a household's factor market participation. However, the extent of the effect of telephones on participation decreases as the estimation controls for households' observed characteristics and factor endowments. Since the use of a telephone is correlated with a household's observed characteristics, it seems plausible that it is also correlated with the unobserved characteristics of households that may be correlated with the market participation of households. So, if a household's unobserved propensity to participate in the market is correlated with telephone use, then the single equation estimate is subject to an omitted variable bias.

If the households that are most likely to use telephones are also those most likely to adopt measures that reduce transaction costs, then the single equation estimation would overestimate the impact. However, it is also possible that households with the highest information/TC use a telephone. In such cases, the use of a telephone might not be able to bring down the transaction costs sufficiently enough to participate in the market. In this case, the single equation model would underestimate the benefit of telephone use.

To address the issue of an omitted variables bias, the estimation allows for the possibility that a household's telephone use and market participation are correlated. Since two variables of interest, use of a telephone and factor market participation, are both discrete, the appropriate simultaneous equation model in this context is a bivariate probit.[126]

A household will use a telephone if the net benefits of telephone use are positive. Similar to market participation given by equation (8.4), the net benefits of telephone use, T_i^*, can be modelled as a latent variable being determined by the following linear equation:

(8.6) $$T_i^* = \eta' Z_i + \varepsilon_{2i}$$

where Z_i is a vector of covariates that capture a household's demand for information and other attributes related to information availability, ε_{2i} is a normally distributed error term(?) with mean zero and unit variance. The conditional probability that for household i given Z_i, one observes $T_i = 1$, (use telephone) is given by

(8.7) $$P(T_i = 1|Z_i) = P(\eta' Z_i + \varepsilon_{2i} \geq 0) = F(\frac{\eta' Z_i}{\sigma_2})$$

To allow for the possibility that unobserved characteristics of a household's decision to participate in the factor market and a household's preference for telephone use are correlated, it is assumed that ε_{1i} and ε_{2i} are distributed according to a bivariate distribution with $E(\varepsilon_{1i}) = E(\varepsilon_{2i}) = 0$, $\text{var}(\varepsilon_{1i}) = \text{var}(\varepsilon_{2i}) = 1$ and $\text{cov}(\varepsilon_{1i}\varepsilon_{2i}) = \rho$. Given that both the decision to participate in the factor market and the decision to use a telephone are dichotomous, the likelihood function is a bivariate normal probit.

Assuming that a telephone is a normal good, the use of a telephone is determined by the price of the telephone and its alternative and a household's income. In addition to that, there are two sources, a household's characteristics and information availability at village/community level that determine the demand for information and hence telephone use. The study has estimated the probabilities of telephone use given Z_i, where Z_i includes variables related to information availability, such as distance to the nearest telephone, and the number of telephone users in a village as a percentage of sample size, as

[126] See Evans et al (1999).

described in section-8.4. The other set of variables that Z_i includes are household characteristics such as income and migration.

Table-8.10 reports the coefficients of determinants of telephone use estimated by standard probit method and table-8.11 reports the coefficients of the market participation estimated by bivariate probit that corrects for the omitted variables bias. Note that all the reported coefficients are normalized probit coefficients. Identification of the bivariate probit model requires that at least one covariate in X_i is not included in Z_i.[127] For this reason, a household's characteristics that have been included in X_i have not been considered in Z_i and vice versa. The covariance between ε_{1i} and ε_{2i}, ρ is equal to -0.0502 with a level of significance of 0.4010. Results reported in table-8.9 and table-8.11 are comparable up to a certain extent. As can be seen from table-8.11, the bivariate probit results are similar to single equation probit estimation; the difference in factor market participation rate between households that use a telephone and households that do not use a telephone is significant and remained at around 8 percent (column three, table-8.11) or more.

Table-8.10: Determinants of Telephone Use: Standard Probit Coefficients (Normalized)

Regressors	Coefficients
Log of household expenditure	0.2893
	(0.0444)**
Distance of nearest telephone in KM	-0.0044
	(0.0068)
Telephone users as a % of sample size	0.01
	(0.0019)**
Family member migrated abroad	0.0581
	(0.0275)*
Family member migrated to city	0.0269
	(0.0378)
Number of observations	284
Log likelihood	-113.28388
LR chi2	121.75
Probability >chi2	0.0000
Pseudo R2	0.3495
Observed probability	0.3028169
Predicted probability	0.2366276

Standard errors are in the parentheses. **, *: Significant at 1%, and 5%.

[127] See Maddala (1983, page 229).

Table-8.11: Factor Market Participation, Bivariate Probit Coefficients (Normalized)

Regressors	Model 1	Model 2	Model 3
User Status (fitted)	0.1726906	0.1362	0.0806
	(0.0782)*	(0.0619)*	(0.0481)~
Household Characteristics:			
Gender, household's head (Male=1)		0.1394	0.0583
		(0.1608)	-0.1248
Age, household's head		0.0111	0.0175
		(0.0074)	(0.0069)*
Age square		-0.0001	-0.0002
		(0.0001)	(0.0001)*
Education (literate=1)		0.0981	0.037
		(0.07)~	(0.0451)
Occupation (Agriculture=0)		0.1002	0.0717
		(0.0385)*	(0.0312)*
Ln of farm size			0.0378
			(0.0114)**
Ln of labour supply			-0.0275
			(0.0101)**
Number of observations	284	284	284
Log likelihood	-119.94062	-85.3576	-72.9388
LR chi2	5.02	74.19	99.02
Probability >chi2	0.0250	0.0000	0.0000
Pseudo R2	0.0205	0.3029	0.4043

Standard errors are in the parentheses. **, *, ~: Significant at 1%, 5%, and 10%, respectively.

8.5.2 *Type of Factor Adjustment, Correcting for Participation*

Once a household participates in factor market(s), does the use of a telephone have any impact on which specific factor market the household participates in? It has been seen in section-8.2 that households that participate in factor markets do not participate in the same way. Among the households that participate in factor markets, in general, 22 percent of them participate in the labour market only, 20 percent of them participate in the land market only, and 13 percent of them participate in both land and labour markets. For the households who have crossed the participation threshold, what are the characteristics that determine the participation in the labour market or in the land market or in both markets? And does the use of a telephone play any role at this stage?

Equation (8.4) describes a household's market participation M_i^* as a function of a household's telephone use and observed characteristics. Similar to that, one can define for each household i, the net benefit of participating in a particular market denoted by K_{ij} as a function of information availability and other characteristics denoted by Y_i, and this can be described by the latent variable model:

$$K_{ij}^* = \gamma Y_i + \varepsilon_{3i} \qquad (8.8)$$

Here $j = 1,2,3$, and stands for land market, labour market, and both land and labour markets and ε_{3i} is a normally distributed error term with mean zero and unit variance, and γ is the vector of parameters to be estimated.

Since participation in a particular market j for household i is preceded by market participation decision M_i^* given by (8.4), one observes K_{ij} only for households for which $M_i^* \geq 0$. Following Heckman (1979), the study corrects for this sample selection bias using the same device in probit analysis.[128] Taking into consideration the issue of sample selection bias, (8.8) can be rewritten for the sub-sample of households for which $M_i^* \geq 0$:

(8.9) $$E(K_{ij} | Y_i, M_i^* \geq 0) = \gamma Y_i + E(\varepsilon_{3i} | Y_i, M_i^* \geq 0)$$

Denoting ρ as the correlation coefficient between ε_{1i} and ε_{3i}:

(8.10) $$E(\varepsilon_{3i} | Y_i, M_i^* \geq o) = \rho\lambda_i$$

where $\lambda_i = f(A_i)/F(-A_i)$, $A_i = -[\beta' X_i]$, and f and F are the standard normal pdf and cdf respectively. Introducing the correction term, equation (8.9) can be rewritten as:[129]

(11) $$K_{ij}^* = \gamma Y_i + \rho\lambda_i + \widetilde{\varepsilon}_{3i}$$

where $E(\widetilde{\varepsilon}_{3i} | M_i^* \geq 0) = 0$ and $E(\widetilde{\varepsilon}_{3i}^2 | M_i^* \geq 0) = \tau_i^2$, and $\tau_i^2 = 1 + \rho^2 \lambda_i (A - \lambda_i)$. Assuming K_{ij} is the observed bivariate state of K_{ij}^*, dividing equation (8.11) by $\tau_i (\tau_i > 0)$, the estimating equation for the sub-sample of households with $M_{ij}^* \geq 0$ is:

(8.12) $$\begin{aligned} K_{ij} &= 0 \; if \; \gamma'(Y_i / \tau_i) + \rho(\lambda_i / \tau_i) + \varepsilon_{4i} < 0 \\ &= 1 \; if \; \gamma'(Y_i / \tau_i) + \rho(\lambda_i / \tau_i) + \varepsilon_{4i} \geq 0 \end{aligned}$$

where $E(\varepsilon_{4i} | M_i^* \geq 0) = 0$ and $E(\varepsilon_{4i}^2 | M_i^* \geq 0) = 1$.

The study has estimated three types of factor market participation under the above sample selection framework. Table-8.12 reports the estimated coefficients. In addition to telephone use status as a regressor, it has included a household's factor endowments and factor prices described in section-8.4.

[128] See van de Ven and van Praag (1981), where a dichotomous choice is used to correct for sample selection bias.

[129] For a complete derivation see Heckman (1979, pp. 156-77). See also van de Ven and van Praag (1981), pp. 235-39.

Under the perfect market assumption where agricultural households act as price takers, both labour-adjustment and land-adjustment decisions depend on wages and rent, and on income. For identification purposes, with the exception of telephone use status, the set of variables that has been included in the selection equation has not been included in any particular factor adjustment decision. For all three cases, the selection equation has included the following variables: gender, age, age square, educational level and occupation of the head of the household, and the household's income.

Table-8.12: Type of Factor Adjustment, Given Market Participation

Covariates	Land Only	Labour Only	Both Land and Labour
Telephone use status (user=1)	0.3456	-0.4585	-0.1359
	(0.3538)	(0.3789)	(0.3208)
Log of land price	0.1965		0.195
	(0.1751)		(0.1561)
Log of wage		-1.8897	0.7553
		(0.6249)*	(0.5514)
Log of initial stock of land	0.0569		-0.1229
	(0.0743)		(0.0619)*
Log of initial stock of labour		0.189	0.0357
		(0.2259)	(0.1845)
Constant	-3.6906	6.793	-3.7963
	(1.7786)*	(2.6806)*	(2.3239)
Covariates for Selection			
Telephone use status (user=1)	0.8436716	0.8496	0.9501
	(0.3528)*	(0.3629)*	(0.351)*
Gender of the head of the HH	0.6541	0.4141	0.6432
	(0.4358)	(0.4345)	(0.4368)
Age of the head of the HH	0.0923	0.099	0.1034
	(0.0415)*	(0.0444)*	(0.0391)*
Age square	-0.0009	-0.0009	-0.001
	(0.0004)*	(0.0005)*	(0.0004)*
Education (Illiterate=0)	0.1796	0.1608	0.0651
	(0.2544)	(0.2869)	(0.2405)
Occupation (Agriculture=1)	0.6307	0.4679	0.716
	(0.1915)**	(0.1967)*	(0.1934)**
Constant	-2.6113	-2.387	-2.7652
	(1.1305)*	(1.1172)*	(1.0651)*
Log likelihood	-233.1928	-234.3181	-233.7675
Wald Chi2	40.50	29.77	67.80
Probability > chi2	0.0000	0.0001	0.0000
Number of observations	284	284	284
Censored observations	44	44	44
LR Test of independent equations (rho=0)			
Chi2	6.89	0.06	13.10
Probability > chi2	0.0086	0.8096	0.0003

Standard errors are in the parentheses. **, * Significant at 1% and 5% respectively.

Telephone use status has significant impact on market participation in general, as can be seen from the selection part of table-8.12. Households that use a telephone are more likely to participate in the market. The signs and coefficients of the selection equation are very much in line with the results of factor market participation that have been derived in the previous subsection.

Table-8.13 reports the marginal effects of telephones on the type of factor adjustment correcting for sample selection. As can be seen from the table, once a household participates in the market, the use of a telephone does not have any significant impact on the type of factor market participation.

Table-8.13: Type of Factor Adjustments: Marginal Effects

Covariates	Land market only	Labour market only	Both land and labour market
Telephone use status (fitted)	0.0957	-0.1366	-0.0534
	(0.098)	(0.124)	(0.126)
Log of land price	0.0544		0.0766
	(0.0487)		(0.0613)
Log of wage		-0.5629	0.2967
		(0.2103)*	(0.2163)
Log of initial stock of land	0.0157		-0.0483
	(0.0205)		(0.0242)*
Log of initial stock of labour		0.0563	0.014
		(0.0656)	(0.0725)

Standard errors are in the parentheses. **, * Significant at 1% and 5% respectively.

8.6 Conclusions

One important policy issue concerning agricultural households in developing countries is their selective factor market participation both for land and labour. One reason that has been proposed in the literature for selective market participation of rural households is the imperfect information and the presence of transaction costs. In this study, in the case of Bangladesh it has been seen that if land and labour markets do not function properly, partly because of costly and imperfect information, increase in information availability in the form of access to a telephone may change the functioning of markets and households' market participation.

Using data from a household survey in Bangladesh, the study has found that use of a telephone might increase the farm households' market participation by as much as 10 to 14 percentage points. However, as has been argued in the chapter, there is a reason to be concerned that this estimate is subject to an omitted variable bias. One possible source is that the households that are most likely to use a telephone are also those most likely to adopt measures to reduce information imperfection and transaction costs. However, correction of this bias does not eliminate the impact of telephone use on factor market participation:

use of a telephone increases factor market participation by at least 8 percentage points even when controlling for possible omitted variable bias.

In addition to the impact of factor market participation in general, the chapter has estimated the impact of telephones on specific factor market participation using a two-stage probit model. The findings of the present estimation suggest that though the use of telephones increases farm households' land and labor market participation, once a household is already in the market, the use of a telephone does not have any impact on which specific factor market the household is going to participate in.

9 MEASURING THE INDIRECT IMPACTS: ACCESS TO INFORMATION AND CHOICE OF MARKETING CHANNELS

9.1 Introduction

While the previous chapter has focused on the impact of telephone on rural households factor market participation, the present chapter focuses on the impact of telephone product market participation. This chapter studies the impact of information cost and other transaction costs on the rural producers' discrete choice between selling to middlemen or direct buyers, and continuous choice of selling intensity to middlemen and direct buyers. More specifically, it tries to address two questions: first, does access to information bring any change in the producers' discrete choice between selling to middlemen vis-à-vis selling to direct buyers? Second, does access to information bring any change in the continuous choice of selling intensity between selling to middlemen vis-à-vis selling to direct buyers? This chapter addresses these two questions under a transaction cost economics framework while controlling for other possible economic factors. It considers access to telecommunications as a proxy for the access to information, and examines the issue of transaction costs for the rural producers of Bangladesh. Transaction cost is a widely used approach to explain the observed market failures and self-sufficiency in agriculture in developing countries.[130]

Transactions can take place directly between buyers and sellers or indirectly through intermediaries. Intermediaries are economic agents who specialize in the activities of buying and selling the same product(s). Better known as middlemen in the context of developing countries, they mediate selling between the seller of a product and its potential buyers. Existence of frictions in trade gives rise to the function of intermediation. In the case of direct transactions, i.e. when a transaction takes place directly between buyers and sellers, they share the transaction (trade) surplus. In the case of middlemen negotiated trade, middlemen share the surplus with buyers and sellers. Economic literature rationalizes the intermediation by arguing that intermediaries emerge because they are able to economize on the cost of transactions and information asymmetries.[131]

One fundamental source of transaction costs is the information cost. Neo-classical economists essentially assume that information is costless and perfect, which does not comply with the reality, particularly in developing countries.[132]

---

[130] See Key et al (2000) for a recent theoretical as well as an empirical analysis on transaction costs.

[131] See Townsend (1978) for cost of transactions, and Freixas and Rochet (1997, p. 16) for information asymmetries in the context of financial intermediaries.

[132] As put in Stiglitz (1988), 'If information transmission and processing were costless and if there were no incentive problems- so everyone transmitted all of his information accurately-

When information is not costless this has important implications in present and future contracts and transactions. Transactions and contracts, which could be feasible in the presence of perfect information, may not occur.[133] In developing countries, the economic agent overcomes this informational problem, whether ex-ante adverse selection, interim moral hazard, or ex-post the costly state of verification, through different informal arrangements and institutions. The existence of middlemen can be viewed as one of the arrangements to overcome transaction costs and costly information.

Access to telecommunications has potentials to overcome limitations associated with information imperfection, and to achieve a reduction in transaction costs. 'Two key determinants of market emergence are the costs associated with acquiring information, and the cost of negotiating transactions'[134], and the spread of telecommunications is expected to reduce both acquiring and negotiation costs. Recent expansion of telecommunications services to rural areas in Bangladesh (described in chapter 4) has increased the access of rural households to information. With this background, the study examines the impact of transaction costs on a rural households' marketing choice between intermediaries and direct marketing where search and information cost is a part of transaction costs.

The study has chosen three agricultural products, eggs, poultry and milk, which are produced by relatively poor rural households in Bangladesh. These products have been used as a source of income generating activities in rural Bangladesh following NGO-led micro credit and poverty alleviation programs. As micro credit usually requires weekly payments, these products generate income streams that fit with the payment requirements of micro credit better than that of other agricultural products, e.g. rice and other staple crops. As a result, both NGOs and their borrowers prefer these products. Although production and consumption separability is an important question for agricultural households, as the production is credit supported, the chosen products are basically produced for the market. In addition, surveyed households that produce these three products participate in the market in one or the other form.

Table-9.1 shows the information source about the product price of rural farm households that have been surveyed in Bangladesh, and table-9.2 shows the discrete choice of marketing channels between selling to direct buyers and middlemen. Both tables compare the observations between villages with a telephone and villages without a telephone. It appears from table-9.1 that for all

clearly there would be everything to be gained, and nothing to be lost, from the centralization of information. But these assumptions are no more realistic than the assumption that goods are costlessly produced, and drop freely from the sky, like manna from Heaven'.

[133] See Akerlof (1970).

[134] See Leff (1984).

three products, producers that live in villages without a telephone rely more on middlemen to get information about the price of their products. Coincidentally, as can be seen from table-9.2, a farm household's discrete choice of marketing channels between middlemen and direct buyers seems also to be different between households living in villages with a telephone from the households living in villages without a telephone. For all three products, the former group of households selects direct buyers over middlemen more often than the latter group. It appears from the observed pattern that the difference in access to telephones can lead to a difference in sources of information and a difference in choices of marketing channels of rural producers.

Table-9.1: Source of Information about Price: Village with and without Telephone[135]

Information	Egg Producers		Poultry Producers		Milk Producers	
Source	A	B	A	B	A	B
Middleman	28.21	4.35	21.21	8.70	5.71	0.00
Not from middlemen	71.79	95.66	78.79	91.3	94.29	100
Total	100.00	100.00	100.00	100.00	100.00	100.00
No of producers	39	23	33	23	35	10

A: Village without Tel; B: Village with telephone.

Table-9.2: Households' Choice of Marketing Channels: Village with and without Telephone

Sell to direct buyers	Village without phone		Village with phone		Total	
No	38	(33.04)	4	(6.90)	42	(24.28)
Yes	77	(66.96)	54	(93.10)	131	(75.72)
Total	115	(100.00)	58	(100.00)	173	(100.00)

Numbers in the parentheses are the percentage.

To explain the observed relationship between access to a telephone and a producer's choice between direct selling vis-à-vis mediated selling, the study uses transaction cost economics (henceforth TCE) developed by Coase as the analytical framework.[136] The usual assumption is that the producer is a cost minimizer and, under a given technology and institutional environment, his/her choice between direct selling versus mediated selling is determined by information cost and other transaction costs minimizing motive. For empirical estimation, it estimates a probit model to examine the impact of transaction cost variables on a producers' discrete choice between mediated vis-à-vis direct

[135] Unless otherwise stated, all the data used in this chapter are from the primary survey described in the Appendix-I. For a brief review of survey coverage and method, see section-9.4.

[136] Coase (1937) in his seminal paper, 'The nature of the firm' first introduced the term 'transaction costs' to explain why firms exist. Later, economists have incorporated transaction cost framework to explain different economic outcomes. Many economists have subsequently extended the TCE framework. See for example Williamson (1975, 1991). See also Fahlbeck (1996) and the references therein (p.1)

selling. For selling intensity, it estimates a two limits tobit model to examine the observed pattern under a transaction costs framework.

The set of articles that are concerned with explaining empirically the existence of middlemen in the context of developing countries is limited mostly to the middlemen's margin on transactions.[137] No previous attempt has been found to explain a producer's choice between middlemen and direct selling incorporating the role of transaction costs. The other distinguishing feature is the choice of products. While previous attempts are mostly concerned with staple crops where production has a high self-consumption motive, products that are considered in this chapter are basically produced for the market. This difference in product choice has an additional rationale. A problem with staple crops is that the observed price gap between buyer and seller can arise not only due to transaction costs but also due to transfer costs. Middlemen usually add some services in between that turn the middlemen's product into something superior compared to raw farm products. All the three chosen products in the context of rural markets in Bangladesh are homogenous and middlemen do not include any explicit transfer value there.

The remainder of this chapter is arranged as follows: section 9.2 describes different origins of transaction costs and formalizes the problem; section 9.3 describes the rural producers and their marketing options; section 9.4 estimates the impact of transaction costs both on the discrete choice of marketing channels and on selling intensity of rural producers; and section 9.5 concludes.

9.2 Transaction Costs and Producers' Choice: An Analytical Framework

Economists have reasoned the existence of middlemen from different perspectives.[138] However, following the literature on financial institutions, the current study rationalizes the functioning of middlemen by arguing that intermediaries emerge because they are able to economize on the cost of transaction.[139] Transaction cost (henceforth TC) is the gap between buying and selling price.[140] Although TC affects both buyers and sellers, the current study limits itself to the impact on sellers.

One standard assumption of neoclassical economics is costless transaction at market clearing prices, which means exchange involves zero transaction cost. However, once one departs from the frictionless world of neoclassical economics, one can find that transaction essentially involves search for

---

[137] See for example, Minten and Kyle (1999).

[138] For example, middlemen reduce the time period that buyers and sellers have to wait for a transaction to take place (Rubinstein and Wolinsky (1987)), they reduce the moral hazard problem otherwise faced by buyers (Biglaiser (1993)), and improve welfare when product quality is not immediately observable by buyers (Li (1998)).

[139] See Townsend (1978) in the context of financial intermediaries.

[140] See Hirshleifer (1984), pp. 421-23.

potential buyers, negotiation with them, and enforcement of transaction/contract. Each of these steps adds costs to the transaction and generates a wedge between the buying and the selling price. As a result, transaction costs reduce the market size and, in extreme cases, when transaction costs are very high, the market may fail.

Transaction costs have various origins. However, following Hobbs (1997), the different origins of TCs can be arranged under three categories: i) information and search costs, ii) negotiation costs, and iii) monitoring and enforcement costs. It may be noted that these categories are neither independent nor mutually exclusive as one could influence another. For the present purpose, all the three categories are discussed in brief.

Information and search costs are an important part of transactions costs (though information problems arise in other contexts as well).[141] The role of information enters into the exchange in the form of communication. One characteristic of a perfect market is perfect communication. For voluntary exchange to take place, offers must be communicated and alternatives from both the buyer's and the seller's side must be compared. Due to the high cost or absence of information, these two essential preconditions of exchange cannot be met and that results in either limited/less than equilibrium or, in extreme cases, complete absence of certain desired transactions.

In addition to collecting information and searching for potential buyers and sellers, economic transactions are usually preceded by negotiations. The negotiation cost of a particular transaction depends on market structure, producer's dependency and product knowledge. While a competitive market provides a producer with more certainty about selling the product, a producer's dependency on a limited number of buyers decreases his/her bargaining power and increases negotiation costs. It follows that a producer's choice of marketing channel changes the negotiation costs as it changes his/her bargaining power. In addition, high dependency on any product and poor knowledge about the product can influence bargaining power negatively; hence a high negotiation cost results. Other sources of negotiation cost include frequency of sale and risk about the product.

Monitoring and enforcement costs are incurred by a buyer/seller in order to monitor the performance of the counter party and to execute contracts. Transaction cost related to monitoring and enforcement costs varies depending on the type of contract between buyer and seller, payment type, and enforcement of contract. Transactions that require repeated performance of the parties involved, i.e. performance related to a transaction that does not end at a

[141] See Dasgupta and Stiglitz (1988).

single point in time but extends over more than one period, usually requires monitoring of one party by the other. Such a type of transaction usually involves higher uncertainty and higher costs. Depending on the legal system, the potential cost to a producer for enforcing a contract can be substantial enough to reduce the incentive of a long-term contract.

It is now possible to formalize the problem by assuming that under a given institutional and technological framework a seller's decision between middlemen versus direct buyer depends on transaction costs minimization. If p is the market price, tc^m and tc^d are the transaction costs associated with selling to middlemen and to direct buyers, and p^m and p^d are the price that a seller gets when he sells to middlemen and direct buyer, respectively, then the marketing choice, y, of a seller for a particular product depends on:

(9.1) $$y = y(p^m, p^d; Z^y)$$

Here Z^y is the institutional environment. Assuming that each producer is a cost minimizer (or equivalently profit maximizer), under a given institutional environment, a producer will opt for direct selling when $p - p^m > p - p^d$. Alternatively she/he will choose middlemen when $p - p^m < p - p^d$ and she/he will be indifferent between middlemen and direct selling when $p - p^m = p - p^d$. However, as both p^m and p^d depend on transaction costs that the producer faces and $p^m = p - tc^m$ and $p^d = p - tc^d$, the seller's decision rule depends on p, tc^m and tc^d:

(9.2) $$y = y(p, tc^d, tc^m; Z^y)$$

Assuming that the market price p is exogenous to the model, then the seller's supply decision, that means the choice between middlemen and direct selling, given by (9.2) can be rewritten as:

(9.3) $$y = y(tc^d, tc^m; Z^y)$$

9.3 Rural Producers, Marketing Options and Prices

The rural economy in Bangladesh is dominated by crop production. As a result, the majority of households living in rural areas are involved in crop production either as landlords, farmers and sharecroppers or as agricultural workers. In contrast to crop production, households that produce eggs, poultry and milk for the market are usually both land and labour poor and have limited access to crop production. Table-9.3 presents some basic characteristics of producers of these three products and compares them with the full sample that does not include these producers (labelled as partial sample) and the full sample that includes them (labelled as full sample).

Table-9.3: Small Producers vis-à-vis Full Sample (mean & standard deviation)

Characteristics	Egg, Poultry and Milk Producers	Partial Sample	Full Sample
Land ownership (in decimal)	93.64 (68.17)	180.17 (286.41)	154.58 (246.19)
Household size	5.17 (1.41)	5.47 (1.86)	5.38 (1.75)
Income, yearly total (in Taka)	47782.50 (32288.15)	86033.08 (124036.53)	74719.53 (106911.4)
Expenditure, yearly total (in Taka)	54634.11 (45198.67)	92778.40 (111474.22)	81496.29 (98190.44)

Numbers in the parentheses are the standard deviations

As can be seen from the table, the mean amount of land that producers of eggs, poultry and milk own is less than that of both the two other samples. Taking household size as a proxy for family labour supply, these households are relatively labour poor compared to the other two samples' households. The claim of their resource poorness is also reflected in their income and expenditure, which are lower than that of their counterparts. In addition, assuming that NGO membership is a source of micro credit for such products promoted by NGOs in Bangladesh, one can see that 38.1 percent of the producers are NGO members compared to 26.5 percent in partial sample and 29.9 percent in full sample.

The different channels that rural producers in Bangladesh utilize in marketing their products are brokers, wholesalers, direct buyers, and bringing to the nearest market. Assuming that there is no difference between selling to the direct buyer and bringing to the nearest market place and selling, i.e. considering both methods as selling to direct consumers, all these different channels can be arranged under two categories: mediated selling and direct selling. Table-9.4 reports the choice of marketing channels of rural producers.

Table-9.4: Market Participation Behavior of Rural Producers and Marketing Channels

Product Name	Percentage of households that sell:		
	Only to middlemen	Only to direct buyers	Both to middlemen and direct buyer
Egg	32.81	34.38	32.81
Poultry	17.54	49.00	33.46
Milk	21.15	36.53	42.32

As seen in table-9.4, the marketing behaviour of agricultural households is essentially censored in nature. Among the 64 households that produce eggs, 32.81 percent of them sell 100 percent of their sales to middlemen, 34.38 percent of them sell 100 percent of their sales to direct consumers, and the remaining 32.81 percent sell both to middlemen and direct consumers. Among the 57 households that produce poultry, 17.54 percent of them sell 100 percent of their sales to middlemen, 49 percent of them sell 100 percent of their sales to direct consumers, and the remaining 33.46 percent sell both to middlemen and direct consumers. Among the 52 households that produce milk, 21.15 percent of them sell 100 percent to middlemen, 36.53 percent sell 100 percent of their sales to direct consumers, and the remaining 42.32 percent sell both to middlemen and direct consumers.

To explore the impact of information in the form of access to a telephone on the choice of marketing channels between mediated and direct selling, the table-9.4 has been rearranged into table-9.5. Table-9.5 reports the selling intensity to direct buyers and compares the producers under two categories: producers from villages with a telephone and producers from villages without a telephone. It appears that for all three products, the former group of producers choose direct buyers over middlemen more than the latter group and the differences between the producers of each of the products are significant.

Table-9.5: Production Sold to the Direct Buyers (Mean): Village with and without a Telephone

Products	Village without a telephone		Village with a telephone		Mean difference
Egg	42.20	(42.69)	80.43	(33.37)	38.24**
Poultry	56.06	(41.75)	96.88	(9.07)	40.81**
Milk	34.10	(43.75)	90.91	(30.15)	56.81**

Numbers in the parentheses are the standard deviation; **Significant at 1% level.

To explore the role of price in selecting marketing channels, the study compares the average selling prices that the two groups of producers received when they have utilized only one of the two channels, i.e. it includes two polar choices only. Table-9.6 reports the average (mean) selling price along with the difference in means. With the exception of eggs, there is no significant difference between the prices received from two channels; it is only the egg producers that sell their products to direct buyers who receive a higher price than those who sell to middlemen.

Table-9.6: Average Selling Price: Mediated vis-à-vis Direct Selling

Price of:	Only to Direct Consumers		Only to Middlemen		Difference in Means
Eggs	11.45	(1.10)	9.40	(1.19)	2.05**
Poultry	68.04	(10.30)	62.00	(12.29)	6.04
Milk	17.05	(4.18)	17.18	(3.16)	-0.13

Standard errors in parentheses; * significant at 5% level; ** significant at 1% level

In addition, the study has calculated the difference between the price in the nearest local market and the actual price that the producers received when they sold to middlemen or to direct consumers. Here the nearest Thana town has been taken as the nearest local market. Thana in Bangladesh is the lowest administrative unit and is characterized by the presence of direct buyers and intermediaries for agricultural products. Table-9.7 reports the difference in average selling prices: the first column is the local market price minus the direct selling price, and the second column is the local market price minus the mediated selling price. As can be seen, none of the differences are statistically significant. This implies that the price that the rural producers receive either from selling to direct consumers or from selling to middlemen is not significantly different from the nearest local market prices. The third column reports the difference of the differences – the difference between the first and second columns; as can be seen, with the exception of poultry, the other two price differences are not statistically significant.

Table-9.7: Difference in Average Selling Price: Local Market vis-à-vis Direct Selling and Local Market vis-à-vis Mediated Selling (Mean and Standard Deviation)

Price of:	Local Market vis-à-vis Direct Consumers		Local Market vis-à-vis Middlemen		Difference in Means~
Eggs	2.14	(2.37)	2.25	(2.12)	-0.107
Poultry	3.57	(4.05)	10.0	(9.72)	-6.4286*
Milk	2.16	(2.81)	2.27	(2.05)	-0.1148

~ Calculated as in table-9.3; standard errors in parentheses; * significant at 5% level; ** significant at 1% level

To relate the difference in price with access to information, one can look at the prices of the products received by the producers of the two categories of villages. Table-9.8 reports the mean price of three different products received by surveyed households and the difference in mean prices. As can be seen from the table, though the mean price of milk and poultry producers from villages with a telephone is higher than their counterparts, none of the price differences is statistically significant. Prices that the middlemen offer do not seem to be statistically different from the prices that the producers receive otherwise from the direct buyers.

Table-9.8: Mean Price Received by Producers and Difference in Prices

Products	Village without telephone		Village with telephone		Mean difference
Eggs	10.32	(1.46)	10.75	(2.54)	-0.433
Poultry	70.88	(14.80)	68.13	(10.30)	2.757
Milk	16.54	(2.98)	18.09	(3.11)	-1.554

Numbers in the parentheses are the standard deviations.

To explore the issue further, a positive relationship between the distance to the nearest telephone and the cost of information have been assumed, and the distance to the nearest telephone has been taken as a proxy for the access to the information. To see the impact of access to information on price differences, all the producers have been divided into two usual categories: producers from villages with a telephone and producers from villages without a telephone. The bivariate correlation shows that the distance to the nearest telephone and the difference in price are positively correlated (the correlation coefficient is 0.232), and the correlation is significant (the level of significance is 0.014).

However, in the case of the producers from the villages with a telephone, the correlation coefficient between price difference and distance to the nearest telephone is not significant.

The explorative type of analysis of this section suggests that the availability of a telephone might have some relationship with the rural producers' choice of marketing channels between mediated vis-à-vis direct selling. In addition, the distance to a telephone might also have some relationship with the observed price differences between the local market and the price that the producers receive. However, the analysis cannot provide any direction of the relationship. In addition, the observed relationship might be a mere correlation and not causation. These issues have been explored further in the next section.

9.4 Empirical Estimation

9.4.1 Empirical Specification

The two observed characteristics of the rural producers: discrete choice between mediated selling vis-à-vis direct selling, and the selling intensity have been separately specified.

Discrete Choice: In terms of selecting marketing channels, there are three groups of producers that can be seen from table-9.1; sell only to middlemen, sell only to direct consumers, and sell to both middlemen and direct consumers. To see the discrete choice between mediated vis-à-vis direct selling, the last two groups have been considered together and two groups have been formed. More specifically, these two groups have been formed based on whether a producer sells to a direct buyer or not.

Econometrically, the specification problem follows a latent regression model:

$$y^* = \beta' tc + \varepsilon \qquad (9.4)$$

Where y^* the latent variable is unobserved. What one observes is a dummy variable y defined by

$$\begin{aligned} y &= 1 \quad \text{if } y^* > 0 \\ y &= 0 \quad \text{otherwise} \end{aligned} \qquad (9.5)$$

Here tc is the vector of transaction costs variables. The probit method has been used to estimate this equation. The likelihood function of this model can be written as:

$$L(\beta, \sigma | y_i, tc_i) = \prod_{y_i=0} \Phi(-\beta' tc_i) \prod_{y_i=1} [1 - \Phi(-\beta' tc_i)] \qquad (9.6)$$

The marginal effects of this model can be written as:[142]

$$\frac{\partial}{\partial tc_{ik}}\Phi(tc_i'\beta) = \phi(tc_i'\beta)\beta_k \qquad (9.7)$$

Selling Intensity: To account for the intensity, a two limits tobit model has been used to specify the behaviour of rural producers. The model choice is motivated by the fact that the observations on dependent variables are censored both at the upper and lower ends. Though there are other alternatives, it was Tobin who first discussed this problem of censored data in the regression context (Tobin, 1958). Later, economists have applied tobit models in different contexts, e.g. household expenditure, labour force participation, to mention a few.[143]

The use of a particular marketing channel by a producer has a maximum limit, L_i^u, a minimum limit, L_i^l, and values in between. Now if q_i^* is the latent variable and q_i is the observed counterpart, the model can be defined as:

$$q_i^* = \beta' tc_i + \varepsilon_i \qquad (9.8)$$

$$\begin{aligned} q_i &= L_i^l \text{ if } q_i \leq L_i^l \\ &= q_i^* \text{ if } L_i^l < q_i^* < L_i^u \\ &= L_i^u \text{ if } q_i^* \leq L_i^u \end{aligned} \qquad (9.9)$$

Here, tc_i is the vector of transaction cost related variables, β' is the vector of estimated parameters, and ε_i is the vector of error terms. Following Maddala (1983), the likelihood function of this model can be written as:

$$L(\beta,\sigma|q_i,tc_i,L_i^l,L_i^u) = \prod_{q_i=L_i^l}\Phi\left(\frac{L_i^l-\beta' tc_i}{\sigma}\right)\prod_{q_i=q_i^*}\frac{1}{\sigma}\phi\left(\frac{q_i-\beta' tc_i}{\sigma}\right)\prod_{q_i=L_i^u}\left[1-\Phi\left(\frac{L_i^u-\beta' tc_i}{\sigma}\right)\right] \qquad (9.10)$$

Denoting $\Phi[(L_i^l-\beta' tc_i)/\sigma]$ and $\Phi[(L_i^u-\beta' tc_i)/\sigma]$ by Φ_i^l and Φ_i^u respectively, with corresponding definitions for ϕ_i^l and ϕ_i^u, the conditional and unconditional expectations of q_i can be written as:[144]

$$E(q_i|L_i^l < q_i^* < L_i^u) = \beta' tc_i + \sigma\frac{\phi_i^l-\phi_i^u}{\Phi_i^u-\Phi_i^l} \qquad (9.11)$$

---

142 See Maddala (1983), pp. 22-23.

143 See Greene (2000), pp. 905-6.

$$E(q_i) = \Phi_i^l L_i^l + \beta' tc_i(\Phi_i^u - \Phi_i^l) + \sigma(\phi_i^l - \phi_i^u) + (1 - \Phi_i^u)L_i^u \quad (9.12)$$

Assuming that ε_i is a continuous random variable with mean 0 and variance σ^2, and $f(\varepsilon|tc) = f(\varepsilon)$, the marginal effects in the two limits tobit model is given by:[145]

$$\frac{\partial E[q|tc_i]}{\partial tc} = \beta \times \Pr ob[L_i^l < q^* < L_i^u] \quad (9.13)$$

9.4.2 Data and Summary Statistics

The study uses the data from the household survey conducted in six different regions of Bangladesh. The survey is described in detail in appendix-1. In the survey, the households were stratified on the basis of ease of access to telecommunications infrastructure where access was defined in terms of distance that needs to be travelled to access telecom services. The survey contained questions on the households' production and marketing behaviour, the choice of marketing channels, and the variables related to transactions costs. In addition, the survey also collected respondents' personal and family characteristics. Table-9.9 provides summary statistics of dependent and transaction costs variables used in this chapter.

The first dependent variable, MKT_1, is the bivariate state between selling to direct buyers vis-à-vis selling to middlemen. Here selling to direct buyers assumes the value one and selling to middlemen assumes zero. The second dependent variable, MKT_2, is the sales to direct consumer as a percentage of total sales.

The first independent variable, the difference between the local market price and the actual price received by the producers denoted by P_DIF has already been discussed in section 3. Two variables related to search and information costs are price discovery cost (INF_DIS) and information availability (INF_SPILL). For price discovery cost, the distance of the nearest publicly accessible telephone in kilometers has been taken as a proxy assuming a positive correlation between distance and cost of information. To capture the information spillover within a village that may reduce search and information cost, the users of a particular village as a percentage of sample size of that village has been taken as a proxy. For this, it has been assumed that while there is a positive intra-village information spillover, there exists no such inter-village information spillover. That information flow within a village is higher than that between villages is based on the assumption that, due to social ties and traditions, villagers share information among themselves through different

[144] See Maddala (1983), pp. 160-61.

[145] For a formal proof, see Greene (200), pp. 909-10.

economic and social interactions.[146] As expected, both information and search costs variables, INF_DIS and INF_SPILL, are highly negatively correlated and the correlation coefficient is significant at 1 percent level. To avoid this colinearty, only the INF_DIS has been included in the estimation.

Table-9.9: Summary Statistics

Variables	Variable Name	Mean	Std. Dev.
Dependent Variables:			
Discrete choice bet. middlemen and direct buyer	MKT_1	0.757	0.430
Sales to direct consumer as a % of total sales	MKT_2	58.688	44.080
Independent Variables:			
Price difference	P_DIF	2.9118	4.3770
Search and Information			
Price discovery cost	INF_DIS	4.249	4.982
Information availability	INF_SPILL	28.884	21.130
Negotiation			
Dependency on product	PRO_DEP	7.686	9.972
Dependency on buyer$^+$	BUY_DEP	2.572	1.483
Sales frequency$^+$	SAL_FRE	4.488	0.813
Market structure	MKT_STR	19.659	12.822
Knowledge about product quality$^+$	KN_QLTY	4.607	0.980
Risk about the product$^+$	PRO_RSK	3.272	1.585
Monitoring and enforcement			
Type of contract with the buyer$^+$	CON_TYPE	2.376	1.556
Payment type$^+$	PAY_TYPE	1.821	1.010
Enforcement of contract$^+$	CON_ENF	2.867	0.988

$^+$ Measured in intervals.

In case of negotiation costs, all the variables, except dependency on product and market structure have been measured in intervals, on a scale of 1 to 5. The dependency on the product (PRO_DEP) measures income from the sale as a percentage of total household income. The dependency on a buyer (BUY_DEP) measures the dependency on a single buyer where dependency is measured from no dependency to high dependency on a scale of 1 to 5 respectively. The sales frequency, (SAL_FRE) is defined as how frequently a producer sells the product and is measured from less frequent 1 to high frequency 5. The market structure (MKT_STR) is the number of available buyers for one of the particular products. The knowledge about product quality (KN_QLTY) is defined as whether a producer is aware of the quality of the product and is measured from no to yes on a scale of 1 to 5 respectively. The risk associated with the product (PRO_RSK) is the risk of damage or spoilage due to non-selling and is measured from no risk 1 to high risk 5.

[146] Among the surveyed households, about 19 percent of them reported their neighbour from the same village as a source of information about prices of all the products on an average, which shows the presence of a strong intra-village information spill over.

In the case of monitoring and enforcement costs, all the variables have been measured in intervals. The type of contract with the buyer (CON_TYPE) is defined as whether there is any contract between buyer and seller and is measured from no contract to clearly specified written contract on a scale of 1 to 5 respectively. The payment type (PAY_TYPE) is the type of payment and is measured from immediate cash payment to uncertain payment on a scale of 1 to 5 respectively. The enforcement of contract (CON_ENF) is the cost required to enforce a contract and is measured as no cost to high cost on a scale of 1 to 5 respectively.

Before proceeding further it should be noted that some of the variables as mentioned above are measured on a scale of 1 to 5.. However, for the present estimation purpose they have been treated as continuous variables. In fact one of the reasons for not considering them as dummy variables is the small size of the present sample.[147]

9.4.3 Estimation Results

Table-9.10 and table-9.11 report the value of the estimated coefficients along with their respective standard errors and marginal effects. Both tables contain the same set of explanatory variables described in summary statistics. Variables that have been dropped for high correlation with other regressors are INF_SPILL, BUY_DEP, MKT_STR, CON_TYPE, and PAY_TYPE. In order to be more tractable, the findings have been discussed under three different types of transactions costs.

Table-9.10: The Discrete Choice between Middlemen vis-à-vis Direct Selling: Probit Coefficients; Dependent variable: middlemen=0, direct selling=1. Method: probit

Regressors	Coefficients	Marginal Effects
P_DIF	-0.0431	-0.0098
	(0.0387)	(0.0088)
INF_DIS	-0.1412	-0.0323
	(0.0272)**	(0.0073)**
PRO_DEP	-0.0372	-0.0085
	(0.0146)*	(0.0034)*
SAL_FRE	0.3591	0.082
	(0.2444)	(0.0534)
KN_QLTY	-0.3756	-0.0858
	(0.1913)*	(0.0429)*
PRO_RSK	0.1656	0.0378
	(0.0891)	(0.0201)
CON_ENF	-0.5893	-0.1346
	(0.1891)**	(0.0444)**
Egg dummy	-0.4461	-0.1086
	(0.3384)	(0.0889)

[147] For a full treatment of this issue, see Robinson (1988) that describes the method of consistent semiparametric regression.

Poultry dummy	1.1255	0.2121
	(0.5710)*	(0.0803)
Constant	3.1088	
	(1.7016)	
Observations	169	169
Log likelihood	-53.392	-53.392
Probability > chi2	0.0000	0.0000
Pseudo R2	0.4298	0.4298
Observed probability		0.7573
Predicted probability		0.8545

Standard errors are in parentheses; * significant at 5% level; ** significant at 1% level

Table-9.11: Continuous Choice of Selling Intensity: Tobit Coefficients

Dependent variable: Percentage of Production Sold to Direct Buyers; Method: Tobit

Regressors	Coefficients	Marginal Effects[1]
P_DIF	-2.065	-0.4347
	(1.5844)	(0.3335)
INF_DIS	-8.7053	-1.8323
	(1.3999)**	(0.2947)**
PRO_DEP	-5.4253	-1.1422
	(0.7538)**	(0.1587)**
SAL_FRE	17.7652	3.7400
	(9.484)~	(1.9966)~
KN_QLTY	-6.3908	-1.3455
	(5.9991)	(1.26297)
PRO_RSK	11.2544	2.3694
	(3.9133)**	(0.8238)**
CON_ENF	-18.8789	-3.9745
	(6.7174)**	(1.4142)**
Egg dummy	-51.9093	-10.9341
	(16.0413)**	(3.3771)**
Poultry dummy	22.1838	4.6395
	(20.0578)	(4.2227)
Constant	132.7862	27.9551
	(64.3057)*	(13.5381)*
No of Observations[2]	169	169
Log likelihood	-392.8642	-392.8642
Probability > chi2	0.0000	0.0000
Pseudo R2	0.1442	0.1442

[1] Conditional on being uncensored; [2] 41 left-censored observations, 60 uncensored observations and 68 right-censored observations. Standard errors are in parentheses; * significant at 5% level; ** significant at 1% level

Search and Information Costs: Search and information cost has a significant impact on the marketing behavior of rural producers. Access to information in the form of access to a telephone has a significant positive impact on the discrete choice of selling to direct buyers and on selling intensity. There exists a negative relation between selling to direct consumers and information costs, and as the search and information costs increase, rural households sell more of their products to middlemen instead of selling to direct buyers.

Negotiation Costs: In the case of negotiation costs, the product dependency has a significant impact both on the discrete choice and on the selling intensity to direct buyers; the more dependent a producer is on a single product, the less the probability of selling to direct buyers. As expected, a high sales frequency has a positive effect on the amount of selling to direct buyers. However, it does not have any impact on the discrete choice of marketing channels. The knowledge of producers about product quality affects the amount of direct selling negatively. It implies that as producers become aware of the quality of their products, their bargaining power relative to middlemen and intermediaries increases and reduces the transaction costs related to negotiation as a result. The other source of bargaining power that has a significant positive impact on the selling intensity is product risk; as risk increases, producers tend to choose direct selling over mediated selling.

Two variables that could not be included here due to high colinearity are market structure and dependency on buyers. As market structure is highly negatively correlated with product dependency, and buyer dependency is highly positively correlated with product dependency (the correlation coefficients are -0.509 and 0.564 respectively and both are significant at 1 percent level), it can be expected that market structure has a positive impact on selling to direct buyers and product dependency has a negative impact on selling to direct buyers.

Monitoring and Enforcement Costs: Transaction costs related to monitoring and enforcement costs have a significant impact on the choice of marketing channels and on selling intensity. Contract enforcement is negatively related with selling to direct buyers; as the cost of contract enforcement increases, producers chose to sell their products to middlemen over direct buyers. Two other variables related to monitoring and enforcement costs that could not be included due to colinerity, contract type and pay type are highly correlated with contract enforcement and product dependency respectively (correlation coefficients are 0.727 and 0.448 respectively and they are significant at 1 percent). Due to this high correlation, one can predict that formal contracts of households with middlemen reduce selling to direct buyers, and payment uncertainty also reduces selling to direct buyers.

9.4.4 Causation or mere Correlation?

The empirical findings described above suggest that access to information in the form of access to a telephone has a significant impact on the rural households' discrete choice of marketing channels and it also has a significant impact on selling intensity as well. Although both estimates imply that access to a telephone significantly reduces transaction costs and increases selling to direct buyers, both estimates assume that the match of a household to the distance to a telephone is exogenous. However, if a household's unobserved propensity to sell to a direct buyer is correlated with the distance to a telephone, then the present estimates will be subject to an omitted variable bias. To address the issue of omitted variable bias, the estimation controls for more observed characteristics. In particular, it controls for local market and market-leading physical infrastructure that can influence direct selling.[148]

To control for other forms of marketing infrastructure other than access to telecommunications, access to the nearest market has been added. For physical infrastructure, the distance of the village from the nearest paved road has been chosen as a proxy. As the study is primarily concerned with the market reaching physical infrastructure, due to the dominance of roads in transportation services in Bangladesh, this is a reasonable choice. For proximity to the physical market place, the distance of the village from the nearest Thana head- quarter has been chosen as a proxy. The rationale behind choosing Thana as a proxy for the local market has been described in section-9.3. However, the distance to the local market is significantly correlated with the distance to the paved road (the correlation coefficient is 0.151 and significant at 5 percent level). To avoid possible colinearity, the distance to the nearest local market (Thana) has been dropped, and only the distance of the village to the nearest paved road has been included.

Table-9.12: Effects of Marketing Infrastructure on Household's Marketing Behaviour

Dependent variable:	Discrete Choice (Direct buyer=1, Middlemen=0)		Selling Intensity to direct buyers as a % of total production	
Regressors	Probit Coefficients	Marginal Effects	Tobit Coefficients	Marginal Effects
Road	-0.2365 (0.0989)*	-0.0511 (0.0206)*	-16.288 (4.0220)**	-3.6902 (0.9112)**
Information	-0.1456 (0.0285)**	-0.0315 (0.0075)**	-8.6343 (1.3866)**	-1.9562 (0.3141)**

Standard errors are in parentheses; * significant at 5% level; ** significant at 1% level

Table-9.12 reports the coefficients and marginal effects of roads and information on the household's marketing choice. Table-9.13 and table-9.14 provide the coefficients and related statistics of other regressors. The estimation results show that physical infrastructure has a significant impact on the rural producers' choice of marketing channels. In addition, as this variable is significantly correlated with the proximity to the nearest market, the local market may also play a significant role. However, though access to roads has a significant impact on the household's marketing behaviour, inclusion of this additional variable does not change the significance of the effects of access to

[148] Controlling for more observed characteristics as a remedy of omitted variables has been discussed in Evans et al (1999).

information on the households' marketing behaviour. The results imply that access to information in the form of access to a telephone has a significant impact on the households' choice of marketing channels between middlemen and direct buyers.

Table-9.13: The Discrete Choice between Middlemen vis-à-vis Direct Selling: Probit Coefficients; Dependent variable: middlemen=0, direct selling=1. Method: probit

Regressors	Coefficients	Marginal Effects
P_DIF	-0.0575	-0.0124
	(0.0411)	(0.0088)
INF_DIS	-0.1456	-0.0315
	(0.0285)**	(0.0075)**
ROAD	-0.2365	-0.0511
	(0.0989)*	(0.0206)*
PRO_DEP	-0.0409	-0.0088
	(0.0152)**	(0.0033)**
SAL_FRE	0.4007	0.0866
	(0.2629)	(0.0534)
KN_QLTY	-0.3249	-0.0702
	(0.1933)~	(0.0417)~
CON_ENF	-0.6162	-0.1332
	(0.1927)**	(0.0431)**
Egg dummy	-0.3996	-0.0918
	(0.3413)	(0.0847)
Poultry dummy	1.1737	0.207
	(0.5991)*	(0.0777)*
Constant	3.8236	
	(1.6828)*	
Observations	169	169
Log likelihood	-52.1182	-52.1182
Probability > chi2	0.0000	0.0000
Pseudo R2	0.4434	0.4434
Observed probability		0.7573
Predicted probability		0.8659

Standard errors are in parentheses; * significant at 5% level; ** significant at 1% level

Table-9.14: Continuous Choice of Selling Intensity: Probit Coefficients
Dependent variable: Percentage of production sold to direct consumers

Regressors	Coefficients	Marginal Effects[1]
P_DIF	-2.8975	-0.6564487
	(1.5684)~	(0.3553357)~
INF_DIS	-8.6343	-1.956163
	(1.3866)**	(0.3141422)**
ROAD	-16.288	-3.69018
	(4.0220)**	(0.911224)**
PRO_DEP	-5.3794	-1.218746
	(0.7280)**	(0.1649255)**
SAL_FRE	15.8861	3.599133
	(9.4524)~	(2.141521)~

KN_QLTY	-3.3305	-.7545441
	(5.8804)	(1.332245)
CON_ENF	-18.1304	-4.107602
	(6.4646)**	(1.464619)**
Egg dummy	-47.4054	-10.7601
	(15.4106)**	(3.49140)**
Poultry dummy	16.9887	3.827046
	(19.6164)	(4.444257)
Constant	192.6541	43.64738
	(61.3582)**	13.9012)**
No of Observations[2]	169	169
Log likelihood	-388.249	-388.249
Probability > chi2	0.0000	0.0000
Pseudo R2	0.1542	0.1542

[1] Conditional on being uncensored; [2] 41 left-censored observations, 60 uncensored observations and 68 right-censored observations. Standard errors are in parentheses; ~significant at 10% level; * significant at 5% level; ** significant at 1% level

9.5 Conclusions

Access to information and transaction costs has an important impact on the rural producers' choice of marketing channels between middlemen and direct buyers of agricultural products in Bangladesh. The empirical findings of this chapter suggest that transaction costs, particularly information cost, has a significant influence on the farm households' discrete choice of marketing channels and continuous choice of selling intensity to direct buyers over middlemen. Apart from information costs, transaction costs related to negotiation, and monitoring and enforcement are also important.

For the small producers of rural regions, access to information appears to be important to link them with direct buyers. Under the given circumstances in Bangladesh, access to information in the form of access to telecommunications can reduce the transaction cost related to search and information costs of small producers significantly, and induce direct selling over mediated selling. However, other market reaching physical infrastructures, for instance paved roads and proximity to physical markets, are also important to link the rural producers with the direct buyers.

Aspects of transaction costs related to negotiation costs and hence bargaining power are either intrinsic with the products or broad and difficult to capture. For example, with the absence of any storage and processing facility, high sales frequency of eggs and milk is a necessary precondition due to the intrinsic nature of those products. As a result, producers of such products would like to sell to a regular buyer. As can be predicted a priori, a higher number of buyers are associated with a higher amount of direct selling, which supports the lower transaction costs linked to a competitive market structure.

The monitoring and enforcements costs are primarily to do with the contractual arrangements and legal framework. Producers seem to link with middlemen with formal contractual arrangements that reduce the likelihood of selling to direct buyers. As expected, formal contracts with middlemen and intermediaries reduce selling to direct buyers.

Findings of the chapter also suggest that the existence of middlemen in rural areas ensures price equality. As the empirical findings suggest, price does not work as a significant incentive to show any discrete shift in producers' choice between mediated vis-à-vis direct selling or in selling intensity. That means, controlling for possible sources of transaction costs, the different channels that the rural producers use in Bangladesh do not result in significant price differences, neither between different channels nor between the local market and the channels.

10 SUMMARY AND CONCLUSIONS

10.1 Research Questions Studied

There are four research questions that have been addressed in this study as mentioned in section 1.2. The first one is the economic and development rationales behind the provision of universal access to telecommunications in the context of developing countries; the second one is the limits of state and market mechanisms and the provision of universal access through alternative mechanisms; the third one is the direct welfare impacts of use of telecommunications services; and the fourth question the study has addressed is the indirect welfare impacts of use of telecommunications services. The study has addressed these four issues under two broad areas: the first two under the institutional aspects of the provision of telecommunications services, and third and fourth issues under the welfare aspects of the use of telecommunications services. The present study has focused on these two aspects in the context of rural areas of Bangladesh and Peru. The empirical findings of the present study are based on household surveys that were conducted in the rural areas of the two study countries.

To address the first question, the present study has relied on the general concept of universal service obligation and proposed that there are strong economic and development rationales that call for universal service provision in rural areas in developing countries. To address the second question, the present study has discussed the limitation of the competitive market to taking into account the positive externalities and proposed how alternative mechanisms can provide universal service and yet keep the benefits of the competitive market environment.

To address the third question, the present study has relied on the theory of welfare changes measurement related to the Hicksian compensating variation and equivalent variation. The study has examined the distribution of compensating variation based on income quartile. In the case of equivalent variation, the study has measured it through the use of the contingent valuation method in a simple referendum setting. To address the fourth question, the study relied on the theory of transaction costs and considered information cost as a part of transaction costs and measured the magnitude of change in households' product and factor market participation due to the change in access to telecommunications.

The first research question is discussed in chapter 2; the second one is discussed in chapters 3, 4, and 5; the third research question is discussed in chapter 6 and chapter 7; and the fourth research question is discussed in chapter 8 and chapter 9.

10.2 Summary of Findings

The first part of the study has focused on the rationales of the provision of telecommunications and has evaluated the traditional and alternative provision mechanisms of telecommunications provision in the context of rural areas of developing countries. Chapter 2 has examined the issue of economic and development rationales of access to telecommunications under the general concept of universal service. The rationales are the presence of positive production and consumption externalities, presence of merit goods, the use of universal service as income redistribution instrument and regional policy instrument. However, despite the strong economic and development rationales that support the provision of universal service, such service cannot be achieved within an unregulated market structure. Due to the very nature of network and the existence of externalities, competitive equilibria do not exist. This implies that the first theorem of welfare economics cannot be applied to the telecommunications sector.

Traditionally the pursuit of universal service in industrialized countries has been achieved under a state monopoly through cross subsidization of local calls for long distance calls and/or high cost rural areas for low cost urban areas. However, the economic cost of such cross-subsidization is the loss of allocative efficiency which results from market distortions. Such cross-subsidy is an inefficient instrument as well. In addition, the geographically targeted subsidy may transfer income from low-income users living in low cost areas to high-income users living in high cost areas.

To overcome the shortcomings of state provision, public policy towards telecommunications has focused on liberalizing the part of the markets where competition is feasible and keeping the residual services under a regulated structure. However, there is a trade-off between competition and universal service provision which is not easy to resolve and competition can lead to an outcome which is not socially desirable. Particularly, competition dries up the scope of cross-subsidy and may result in ‘too much’ entry in the profitable part of the market and ‘too little’ entry in the loss making regions.

In contrast to industrialized countries that provide universal service at household level, the goal in developing countries is to provide access to telecommunications at a ‘reasonable level’. As a result, developing countries usually define universal service based on the population of a community, the distance that needs to be travelled to reach the nearest public telephone, and travelling time. There are competing approaches that are practiced in developing countries such as cross-subsidy, cost of service approach and targeted subsidy. However, under these approaches the goal of universal access for the households that live in rural areas of developing countries seems unattainable.

Chapter 3 has examined the state universal service in telecommunications in the rural areas of Bangladesh and Peru including the state of the ICT sector, and the market and regulation. The review has revealed that the development of information and communication technologies in Bangladesh has remained at a very low level. The inadequacy is reflected in terms of network availability, the state of diffusion and adoption of information technologies and in the state of creation of new technologies. A regional level comparison has revealed that it is not only the low-income that prevents Bangladesh from having a larger network, but there are some other factors, for instance, institutional arrangements and regulatory structure that act as constraints.

Unlike Bangladesh, the review for Peru has revealed that the country shows a modest achievement in network development, the state of diffusion and adoption of information technologies, and the state of creation of new technologies. A regional level comparison has revealed that the transformation of the telecommunication sector from a state monopoly to a private monopoly and then to the competitive market has helped to achieve a rapid network growth, which is higher than the regional average.

The review has also revealed that the two countries are characterized both by similarities, such as low penetration in rural areas, and differences such as the state monopoly in fixed telephony in Bangladesh versus the competitive market in fixed telephony in Peru. However, despite the difference in the extent of market openness between Bangladesh and Peru, the universal access has not been achieved either in Bangladesh or in Peru. Particularly, households living in rural areas have not benefited either under a monopoly market structure in Bangladesh or under a competitive market structure in Peru.

Chapter 4 has examined the issue of universal service under market and state ownership and reviewed two alternative provision mechanisms. Though extension of telecommunications to rural areas generates positive consumer surplus, a competitive market may not provide such services, particularly in rural areas, due to the ineffective demand. However, the non-existence of competitive provision does not ask for state provision. As experiences in both Bangladesh and Peru have shown, the government's first level involvement (production of service) is not welfare improving as such provision falls short of universal access and distorts market outcome. Two alternative mechanisms that have been discussed in this chapter are public-private partnership in Peru and business-NGO partnership in Bangladesh; two examples of the use of the market in achieving universal access for rural people in these two countries.

The first example has revealed how government intervention can solve the universal service obligation with minimum distortion. The regulatory authority in Peru has addressed the issue of universal access in rural areas through private provision where private sector participation is ensured through the auction for universal service. Findings have shown that when it is necessary to compensate for universal service for areas where private providers do not voluntarily extend services, auction offers a suitable alternative.

The second example has revealed how universal service can still be achieved when there is no government intervention. The example of business-NGO partnership between Grameen Bank and Grameen Phone revealed that through the complementary inputs from business and NGO, it is possible to devise a provision mechanism that provides universal access without requiring any public transfer or cross-subsidy.

Empirical estimation on the effect of availability of public telephones at village level revealed that in the case of rural households, both in Bangladesh and in Peru, the availability of a telephone at village level has significant positive impact on households' telephone use.

While the first part of the study has focused on the rationales of the provision of telecommunications and the evaluation of institutional mechanisms necessary to provide telecommunications, the second part of the study has evaluated the benefits of such provision focused on the impact of the use of telecommunications. Chapter 5 has sketched the conceptual background for the welfare measurement that is carried out in subsequent chapters. Rural households can be treated both as consumption and production units, and such households use telephones both as a consumption good and as a production input. As a result, the direct welfare impact of the availability of a telephone can be measured by measuring the change in utility that households derived from the consumption of telephone services. The indirect welfare impact of the use of a telephone can be measured by measuring the impact on households' factor and product market participation.

Given the availability, rural households consume telecommunications services as opposed to traditional alternatives and derive utility from the consumption of such services. So the welfare impact of change in access to telecommunications is the change in households' utility resulting from change in access to telecommunications services. For a measurable indicator of welfare, the monetary measures available are: consumer surplus, compensating variation, and equivalent variation.

The indirect impact of a telephone on rural households' production behaviour can be measured from the households' supply response in the form of market participation under the presence of transaction costs. The change in access to a telephone in rural areas may reduce transaction costs that households in those areas face both as buyers and as sellers. Given that a telephone has an impact on the reduction of transaction costs, such reduction would be reflected in the rural households' market participation behaviour.

Chapter 6 has measured the welfare implications through the measurement of Hicksian compensating variation. As Hicksian compensating variation does not suffer from path dependency, as a measure of welfare change, it has important advantages over Marshallian consumer surplus. In addition, Hicksian compensating variation, henceforth CV, permits the exact measurement of the monetary benefits associated with any technological change.

The estimation of CV in the case of telecommunications is complicated by the fact that, due to the regulatory nature of the industry, the direct price of telecommunications does not vary within a country. To circumvent this fact, the chapter has utilized individual time-price-variation to estimate the parameters necessary to calculate CVs. The estimated CVs show that for all four quartiles of income, households derive a positive CV stemming from the availability of telecommunications in rural areas in Bangladesh and Peru.

Chapter 7 has measured the direct welfare impact of a telephone on rural households through measuring the Hicksian equivalent variation. To elicit a households' equivalent variation the chapter has relied on a contingent valuation method and used a simple referendum method in a dichotomous setting to reveal the households' willingness to pay. The use of a referendum method in contingent valuation has become a presumptive approach, as a referendum method appears to be incentive compatible and thus does not suffer from strategic behaviour in revealing the truth.

The estimation of willingness to pay can be done through parametric or non-parametric approximations. The parametric estimation specifies a parametric model in which the relation between the willingness to pay and a household's intrinsic characteristics is stated. In contrast, non-parametric approximation estimates the distribution of the willingness to pay without assuming any parametric specification of the preferences distribution.

The chapter has utilized both parametric and non-parametric methods in estimating rural households' willingness to pay for access to a public telephone. For parametric estimation, the chapter has relied on a logit model and a bounded probit model. For non-parametric estimation, the chapter has relied on Turnbull's algorithm and Kriström's non-parametric measure. Despite differences in willingness to pay figures that are derived from the parametric and non-parametric methods, both parametric and non-parametric estimations suggest that rural telecommunications projects are welfare enhancing.

Chapter 8 has examined the impact of access to information in the form of access to a telephone on factor market participation of farm households in Bangladesh. It has examined whether access to information has brought any discrete change in farm households' land and labour market participation decisions and, correcting for market participation, whether access to telecommunications services has had any impact on the specific type of factor adjustment. The chapter has examined these questions under the transaction costs framework where information cost has been considered as a part of the transaction costs.

The observed behaviour of farm households in Bangladesh has revealed that such households show selective factor market participation. The mean participation rate between households that use a telephone and households that do not use a telephone is significantly different; households that use a telephone have a higher rate of market participation than the households that do not use a telephone. The parametric estimation that has controlled for other observable factors also found that telephone use has a sizable impact on households' factor market participation.

Though the impact of a telephone meets with the prior expectation, there is a reason to be concerned that this estimate is subject to an omitted variable bias. One possible source is that the households that are most likely to use a telephone are also those most likely to adopt measures to reduce information imperfection and transaction costs. However, correction of this bias does not eliminate the impact of telephone use on factor market participation and the impact of telephone use remains a significant factor in farm households' factor market participation decisions.

In contrast to the above findings, the use of a telephone does not have any impact on households' specific factor market participation. The parametric estimation that has corrected for possible selection bias revealed that though use of a telephone increases farm households' land and labor market participation, once a household is already in the market, the use of a telephone does not have any impact on which specific factor market the household is going to participate in.

Chapter 9 has examined the impact of availability of a public telephone at village level on a small producer's marketing choice between direct buyers and middlemen in the transaction cost framework. It has examined whether access to information has brought any change in a rural producer's discrete choice between selling to middlemen vis-à-vis selling to direct buyers, and whether access to information has brought any change in the continuous choice of selling intensity between selling to middlemen vis-à-vis selling to direct buyers.

In analysing the impact of availability of telephones, the chapter has considered information costs and other forms of transaction costs into account.

The observed behaviour of small farm producers in Bangladesh has revealed that the availability of telecommunications services has made a difference in the source of information about market price; while producers from villages without telecommunications services have relied more on middlemen for price information, producers from villages with a telephone have shown less dependence on middlemen. Coincidently, producers from villages with a telephone have opted more often for direct buyers over middlemen compared to the producers from villages without a telephone. The observed pattern has shown that the difference in availability of telephones may lead to a difference in the choice of marketing channels.

The parametric estimation has revealed that the access to information in the form of access to a telephone has significant impact on choice of marketing channels and of selling intensity of rural producers; the lower the information cost, the higher the probability of selling to direct buyers and the higher the selling intensity to direct buyers. The other forms of transaction costs such as negotiation, and monitoring and enforcement costs can also influence the choice of marketing channel as well as intensity.

However, there were reasons to be concerned that the estimates have been subject to an omitted variable bias. Inclusion of market leading physical infrastructure has shown that such factors can have significant influence on a small producer's choice of marketing channels as well. However, inclusion of such covariates has not changed the significance of access to a telephone on the choice of marketing channels.

10.3 Overall Conclusions

Telecommunications services are characterized by the presence of positive production and consumption externalities that demand the provision of universal access. However, due to the very nature of network and the existence of externalities, competitive equilibria do not exist and, due to the allocative inefficiency, state provision is not an optimum solution either. Use of innovative mechanisms, for instance auction for universal service, and business-NGO partnership, can provide alternative solutions for the rural households in developing countries.

The project in Peru that allocates the universal access fund through auction requires a subsidy of US$11 per inhabitant, and for one dollar of subsidy, the project generates two dollars of private investment. However, the source of the subsidy is the ICT sector and the fund generated from the sector is used within the sector to extend the sector to the rural areas of the country. In contrast, the project in Bangladesh does not require any subsidy. In addition, the availability of a public telephone at village level increases rural households' telephone use by around 12 percent in Peru and around 26 percent in Bangladesh.

The estimated compensating variations for the households in Bangladesh vary from US$0.11 to US$1.59 per call, and for Peru the figures vary from US$1.45 to US$2.91 per call. In addition, for Peru, households' willingness to pay (WTP) for a local call varies from US$0.25 to US$0.35, and for a LDN call varies from US$0.33 to US$0.45. For Bangladesh, households' WTP for a local call varies from US$0.10 to US$0.26, for a LDN call from US$0.23 to US$0.50, and for an international call from US$0.93 to US$1.35. Despite the fact that the monetary measures of welfare vary depending on income quartile, household's characteristics, and measurement methods such as parametric and non-parametric methods, figures from all the measures utilized here suggest that the rural telecommunications projects in Bangladesh and Peru are directly contributing to the improvement of welfare of rural households.

If land and labour markets do not function properly partly because of costly and imperfect information, increase in information availability in the form of access to a telephone may change the functioning of markets and a household's factor market participation. In the case of Bangladesh, the difference in factor market participation rate between households that use a telephone and households that do not use a telephone is around 8 percent and the difference is statistically significant.

In addition to factor market participation, access to information and other forms of transaction costs can significantly influence the intermediation process between rural producers and buyers. The decrease in information costs may reduce the role that intermediaries play between small producers and buyers in rural areas. That means, the small producers' access to information may lead to the establishment of a direct link between themselves and the buyers.

10.4 Contributions

The general analytical focus of this study has been on institutional mechanisms of ICT provision and on consumer welfare change measurement in empirical applications. However, in addition to empirical applications, the study has also incorporated economic theory where necessary to avoid probable theoretical ambiguity and interpretational inconsistency.

One of the contributions of the current study has come from describing the economics of alternative provision mechanisms of universal access. The study has identified the factors for which a competitive market structure does not provide such service in rural areas in developing countries. It has also provided examples of two alternatives that could be useful in the context of developing countries. The study has described the operational and functioning mechanisms of these two alternatives and evaluated them in terms of attaining a universal access goal.

The current study has also contributed to measuring the willingness to pay of rural households for access to a public telephone where the study has made use of both parametric and nonparametric measures for the estimation of willingness to pay. Use of the willingness to pay in project benefits measurement is sometimes the only source of measuring the welfare of the project. The measurement has provided the magnitude of the welfare benefits of such projects and can be used for rural telecommunications project decisions.

Another contribution of the study is examining the impact of information on farm households' product and factor market participation. While the critical role of information as a cause of market imperfection has become axiomatic, the theory is backed by little empirical evidence on the magnitude of the problem. The present study has provided a measurement of the impact of information in the rural context.

In the case of the factor market, the study has contributed to the existing literature in two ways. First, the study has empirically estimated the impact of access to information on farm households' factor market participation behaviour. In doing so, the study has controlled for possible self-selection bias by jointly estimating both the propensity of a household to participate in the market and the match of households to telephone use. Second, the study has empirically estimated the impact of access to information on specific factor adjustment decisions. At this stage, the study has estimated a farm household's probability to participate in a specific factor market correcting for participation.

In the case of the product market, the study has contributed to the understanding of the role that intermediaries in rural areas play, and how the availability of information can lead to the establishing of a direct link between small farm producers and buyers in rural areas. In this context, the study has explained the influences of different sources of transaction costs in general and contribution of information costs in particular.

10.5 Future Research

In the case of institutional aspects, there are three areas that could be further explored: first, the political economy of telecommunications market liberalization; second, identification of the factors that contribute to the success of partnerships; and third, identification of areas where such partnerships could contribute.

The first research question would be 'what are the political economy factors that determine the ICTs market liberalisation decision?' Despite the common consensus that state provision of ICTs is inefficient and such provision does not improve distributional equity, not all countries have privatised and liberalised their market. It is not only the economic factors that dictate the privatisation process and determine liberalization outcomes, there are also other political economy factors that play a significant role. It would be important to analyse such factors along with economic factors that determine the liberalization process.

Another research question would be 'what are the factors that contribute to the success of the partnerships described in this study?' The two partnerships described in the study are similar in nature in the sense that they use market mechanism in achieving universal access that market alone does not provide in the specific context of the study countries. Apart from the use of market, the partnerships are different in the sense that one is a government initiative and the other is a non-government initiative. However, both the partnerships are successful in achieving universal access. As a result, it would be interesting to evaluate the factors that determine the success of these partnerships.

The third important research question would be 'what are the other possible areas where such partnerships could contribute?' Apart from the case study described with regard to Bangladesh, the NGO-Business partnership has many more potential applications. For instance, one further research interest in the present context could be the content creation. The present World Wide Web (WWW) acts as a cost efficient source of ideas and resources. However, it is lagging behind in offering resources in the local context of developing countries, and one probable application of business-NGO partnership would be local level content creation. It would be an interesting research idea to examine a possible partnership in this area.

In the case of welfare aspects, some of the areas that could be further explored are: first, impact of telecommunications on labour mobility and wages; second, changes in production composition and market outreach; third, changes in a household's income due to change in ICT use.

Though information availability decreases labour market imperfections and increases a households' labour market participation estimated in the present study, further gain in resource allocation could be achieved through labour mobility across regions within a country. The agricultural wage rate both in Bangladesh and in Peru varies across regions, and in some cases the difference in wage rates is significant. Availability of telecommunications could bring information in wage differentials and could trigger labour mobility. It would be interesting to look at whether wages in different regions are converging and whether labour mobility across regions is a visible phenomenon following the introduction of telecommunications in rural areas.

The second research question would be to evaluate the impact of a telephone on a household's production composition and market outreach. As the availability of telecommunications increases a household's product market participation, one can expect that such participation would lead to a change in production composition, at least in the medium to long run. Related to production composition, access to telecommunications services provides a greater market outreach for households that would help to reap a higher price for the outputs and a lower price for the inputs.

Another future research area could be to estimate the change in a household's income attributable to the use of telecommunications services. It should be mentioned that a straightforward estimation of the impact of telephone use on household income suffers from reverse causality as use of a telephone increases income and increased income contributes to higher telephone use. So one possibility would be to build a simultaneous equation model free of simultaneity and spurious correlation.

The future research ideas described above are limited to the provision and use of telecommunications, and have not included aspects related to information technologies. One immediate research agenda would be to find out mechanisms to provide access for rural people in developing countries to information technologies. Such access could be shared in nature, for instance, in public places. In the case of places where households are using information technologies, the evaluation of the impact at household level would shed important light on the welfare impacts of such technologies.

APPENDICES

APPENDIX-I

The present research uses data from two household surveys conducted in the six different rural regions in Bangladesh and the south rural region of Peru. For both countries, surveys followed a two-stage random sampling procedure in selecting households and contained questions on households' demographic and housing characteristics, employment, income and expenditure, participation in social programs, and availability of infrastructure, access, use and willingness to pay for a telephone, among other factors. Both surveys included settlements with and without a public telephone where a lack of a public telephone service was primarily due to a supply constraint instead of one of demand.[147]

SURVEY DESIGN FOR BANGLADESH

In the case of Bangladesh, the survey was conducted during January and February of 2001 in six different regions consisting of three villages from each of the regions with a sample size of 284 households. The survey followed a two stage stratified random sampling process; at the first stage selecting the strata, and at the second stage selecting the households from each stratum. The survey assumed each administrative district as a region, considered each region as a cluster and arranged all the clusters according to the availability of access to telecommunications. At this stage, the survey selected six clusters randomly. Once the clusters were decided, the next step was to select households from each cluster. In order to improve the efficiency of statistical inference, a stratified random sample design was chosen: the households were stratified on the basis of ease of access to telecom services. Access was defined in terms of distance that needs to be traveled to access telecom services. Thus, three strata were created where a village represented a stratum, and from each village an equal number of households was sampled. Households were selected using the systematic random sampling procedure.

In addition, in the case of Bangladesh, the present study uses information from two other related surveys, ZEF-JU 1998 and ZEF-JU 2000, which were conducted in Bangladesh. The details of these surveys can be found in Bayes et al (1999) and Bayes et al (forthcoming).

The survey consisted of the following sections:

i. Household characteristics;
ii. Housing characteristics;
iii. Income and wealth;
iv. Expenditures;
v. Employment and migration;
vi. Production, market and trade;

[147] Many of the settlement inhabitants without access to public telephone service travel many kilometers to arrive to a settlement that has the service.

vii. Transactions and transaction costs and market participation;
viii. General economic and social conditions;
ix. Availability of infrastructure;
x. Access to and use of telephone;
xi. Willingness to pay for access to public telephone.

Sample Framework: The list of villages provided by the local Grameen Bank branch office has been used as a sample framework.

Sampling Type: Two-stage probabilistic sampling procedure; selecting the rural settlements at the first stage based on the ease of access to a public telephone, and selecting households from selected rural settlements at the second stage.

Sample Size Determination: The sample was determined based on the following formula given by (A.1):

$$n = \frac{p * q * (z)^2}{(e)^2} \qquad \text{(A.1)}$$

Where, n = Size of sample
z = Values from standardized normal table
p = Proportion
q = (1-p)
e = Allowable error.

Based on the formula given by (A.1), the sample size for the Bangladesh survey was determined considering a confidence interval of 95% and a sample error of +/-5.8%:

$$n_{Bangladesh} = \frac{(0.5)*(0.5)*(1.96)^2}{(0.058)^2} = 284$$

So the desired sample size for Bangladesh was 284.

Selection Procedures: The sample was selected based on a two stage stratified random sampling procedure: at the first stage selecting the settlements based on simple random sampling, and at the second stage, selecting the households from each settlement based on systematic random sampling.

Selected Rural Settlements: Out of six clusters, three were with a telephone and three were without a telephone. To increase the efficiency of data, rural settlements were selected based on the ease of publicly accessible telephones. Note that, unless otherwise stated, the publicly accessible telephone in the case of the present survey context in Bangladesh means the village pay phone of the

Grameen Bank village pay phone program. The tables below provide the selected rural settlements in Bangladesh.

Table-A.1.1: Selected Rural Settlements with Telephone Service in Bangladesh

District	Thana	Village	Type[a]
Dhaka	Dhamrai	Bari Gang	1
Feni	Chagal Naia	Shilua	1
Comilla	Burichong	Simail Khara	1
Dhaka	Dhamrai	Burakair	2
Feni	Feni	Sahapur	2
Comilla	Burichong	Sholonoll	2
Comilla	Burichong	Chay Gharia	3
Dhaka	Dhamrai	Ford Nagar	3
Feni	Feni	Satmoti	3

[a] 1=Settlement with public phone; 2= settlement that has public phone within a distance of 3 kilometers; 3= settlement that has public phone within a distance of 5 kilometers

Table-A.1.2: Selected Rural Settlements without Telephone Service in Bangladesh

District	Thana	Village	Type[b]
Manikgonj	Mankigonj	Alinagar	4
Bogra	Bogra	Bashkuta	4
Tangail	Mirzapur	Najirpara	4
Bogra	Bogra	Atail	5
Tangail	Mirzapur	Kudalia	5
Manikgonj	Mankigonj	Thatanga	5
Manikgonj	Saturia	Chartilli	6
Bogra	Bogra	Palaher	6
Tangail	Mirzapur	Salimnagar	6

[b] 4=No public phone at present but has been selected for future extension. 5=settlement that is within 3 km distance of future public phone extension point (4); 6= settlement that is within 5 km distance of future public phone extension point (4).

SURVEY DESIGN FOR PERU

In the case of Peru, the survey was conducted between December 4th and 22nd of 1999 in four rural zones of south Peru: Apurimac, Cusco, Puno and Arequipa. Similar to the Bangladesh survey, the households were chosen randomly by a two-step sample of conglomerates.[148] The survey was conducted on a total of 1000 rural households that included 500 households that belonged to the settlements without a telephone and the remaining 500 households that belonged to the settlements with a telephone.

The survey consisted of the following sections:

i. Household characteristics;
ii. Housing characteristics;

[148] The conglomerates are the settlements with a population between 300 and 3000 inhabitants.

iii. Employment and income of households;
iv. Households' expenditure, durable goods inventories and participation in social programs;
v. Agriculture activities;
vi. Availability of infrastructure
vii. Access and use of telephone
viii. Use and availability of information and communication;
ix. Households' willingness to pay for access to public telephone.

Sample Framework: For the sample framework, the Rural Settlements directory provided by the Telecommunications Investments Fund (FITEL) was used.

Sampling Type: The type of sampling employed is one of two-staged conglomerates. The primary unit of sampling is composed of rural settlements, which were divided ino those with access to a telephone service and those without. Also, settlements were divided according to the type of road (asphalted or saddle path). The second sampling stage refers to the houses of selected rural settlements.

Sample size determination: In the case of Peru, the sample size was determined considering a confidence interval of 95% and a sample error of +/- 3.1%. Using the formula given in (A.1), the size of the sample for Peru is:

$$n_{Peru} = \frac{(0.5)*(0.5)*(1.96)^2}{(0.031)^2} = 1000$$

Selection Procedures: For the selection of rural settlements, a systematic selection process was used.

$$K = \frac{N}{n}$$

Where, K = sampling ratio;
N = Total rural settlements;
N = Size of the sample.

Table-A.1.3: Selected Rural Settlements without Telephone Service in Peru

Department	Province	District	Rural Settlement
Apurimac	Abancay	Curahuasi	Pisonaypata
Apurimac	Abamcay	San Pedro de Cahora	Asil
Apurimac	Andahuaylas	Huancarama	Union San José de Arcahua
Apurimac	Andahuaylas	Kishuara	Tintay
Apurimac	Andahuaylas	Pacucha	Nanchaybamba

Department	Province	District	Rural Settlement
Apurimac	Andahuaylas	Turpo	S.B. Anta
Apurimac	Antabamba	Oropesa	Totora
Apurimac	Grau	Curpahuasi	Tamboraccay
Apurimac	Aymaraes	Colcabamba	Colcabamba
Apurimac	Aymaraes	Chapimarca	Santa Rosa
Apurimac	Cotamanba	Cotabamba	San Juan
Apurimac	Cotabamba	Mara	Pitic
Apurimac	Chincheros	Ocobamba	Challhuani
Apurimac	Chincheros	Cocharcas	Cocharcas
Arequipa	Arequipa	Arequipa	El Chiral
Arequipa	Caylloma	Lluta	Taya
Arequipa	Caylloma	Tisco	Cota Cota
Arequipa	Caylloma	Chivay	Cano Cota
Arequipa	Caravelí	Acari	Molino
Arequipa	Caraveli	Quicacha	Quicacha
Arequipa	La Unión	Alca	Huillac
Arequipa	La Unión	Tomepampa	Locrahuanca
Arequipa	Castilla	Aplao	Cosos
Arequipa	Castilla	Andagua	San Francisco de Tauca
Cusco	Acomayo	Acos	Corma
Cusco	Acomayo	Pomacanchis	San Juan
Cusco	Anta	Chinchaypujio	Pantipata
Cusco	Anta	Limatambo	Tomacaya
Cusco	Anta	Huaracondo	Rahuanqui
Cusco	Urubamba	Maras	Mahuaypampa
Cusco	Calca	Taray	Huancalle
Cusco	Calca	Lamay	Huama
Cusco	Calca	San Salvador	Ccamahuara
Cusco	Paucartambo	Colquepata	Accha
Cusco	Canas	Layo	Cconchupata
Cusco	Canchis	Marangani	Huiscachani
Cusco	Canchis	San Pedro	Raqchi
Cusco	Quipicanchis	Quiquijana	Ttio
Cusco	Chumbivilcas	Colquemarca	Armiri
Cusco	Chumbivilca	Santo Tomás	Llique
Cusco	Paruro	Paccaritambo	Huaninpampa
Cusco	Paruro	Paruro	Mayunbamba
Cusco	Quispicanchis	Urcos	Hurara Huara
Cusco	Quispicanchi	Lucre	Huancarpay
Puno	Lampa	Vila Vila	Vila Vila
Puno	Lampa	Pucará	Chijnaya
Puno	Azangaro	Tirapata	Jurinsaya Aniago
Puno	Melgar	Orurillo	Choquesani
Puno	Huancané	Inchupalla	Inchupalla
Puno	Azangaro	Azangaro	Huayrapata I

Selected Rural Settlements with Telephone Services: For the selection of the rural settlements with telephone services, the distances among them and also the selected rural settlements without a telephone service was considered.

Table-A.1.4: The Selected Rural Settlements with Telephone Service in Peru

Departament	Province	District	Rural Settlement
Apurimac	Abancay	Curahuasi	Curahuasi
Apurimac	Abamcay	San Pedro de Cahora	San Pedro de Cachora
Apurimac	Andahuaylas	Huancarama	Huancarama
Apurimac	Andahuaylas	Kishuara	Kishuara
Apurimac	Andahuaylas	Pacucha	Pacucha
Apurimac	Andahuaylas	Turpo	Turpo
Apurimac	Antabamba	Oropesa	Mamara
Apurimac	Grau	Curpahuasi	Curpahuasi
Apurimac	Aymaraes	Colcabamba	Chalhuanca
Apurimac	Aymaraes	Chapimarca	Tintay
Apurimac	Cotamanba	Cotabamba	Cotabamba
Apurimac	Cotabamba	Mara	Mara
Apurimac	Chincheros	Ocobamba	Ocobamba
Apurimac	Chincheros	Cocharcas	Uripa
Arequipa	Arequipa	Arequipa	Chiguata
Arequipa	Caylloma	Lluta	Lluta
Arequipa	Caylloma	Tisco	Callalli
Arequipa	Caylloma	Chivay	Chivay
Arequipa	Caravelí	Acari	Acari
Arequipa	Caraveli	Quicacha	Chaparra
Arequipa	La Unión	Alca	Alca
Arequipa	La Unión	Tomepampa	Tomepampa
Arequipa	Castilla	Aplao	Aplao
Arequipa	Castilla	Andagua	Andagua
Cusco	Acomayo	Acos	Acomayo
Cusco	Acomayo	Pomacanchis	Pomacanchis
Cusco	Anta	Chinchaypujio	Ancahuasi
Cusco	Anta	Limatambo	Limatambo
Cusco	Anta	Huaracondo	Huarocondo
Cusco	Urubamba	Maras	Maras
Cusco	Calca	Taray	Pisac
Cusco	Calca	Lamay	Lamay
Cusco	Calca	San Salvador	San Salvador
Cusco	Paucartambo	Colquepata	Huancarani
Cusco	Canas	Layo	Layo
Cusco	Canchis	Marangani	Marangani
Cusco	Canchis	San Pedro	San Pedro
Cusco	Quipicanchis	Quiquijana	Quiquijana
Cusco	Chumbivilcas	Colquemarca	Colquemarca
Cusco	Chumbivilca	Santo Tomás	Santo Tomás
Cusco	Paruro	Paccaritambo	Paccaritambo
Cusco	Paruro	Paruro	Paruro
Cusco	Quispicanchis	Urcos	Ccatcca
Cusco	Quispicanchi	Lucre	Lucre
Puno	Lampa	Vila Vila	Palca
Puno	Lampa	Pucará	Pucará
Puno	Azangaro	Tirapata	Tirapata
Puno	Melgar	Orurillo	Orurillo

Departament	Province	District	Rural Settlement
Puno	Huancané	Inchupalla	Putina
Puno	Azangaro	Azangaro	Azangaro

Note: Rural settlements with telephone services located in a distance greater than 4 hours from the rural settlement without telephone access have not been considered.

Replacement Sample: A sampling with replacement (RS) has been considered. This sampling is detailed in the following table:

Table-A.1.5: Replacement Sample Plan in Peru

Province	District	RS with Telephone	RS without Telephone
Lampa	Lampa	Lampa	Huayta central
Azangaro	Asillo	Progreso	Jila central
Melgar	Umachiri	Ayaviri	Umachiri
Huancane	Huancane	Huancane	Munaypa
Andahuaylas	Kishuara	Matapuquio	Vischingay
Andahuaylas	Sta. Maria de chicmo	Moyobamba baja	Taramba
Abancay	Abancay	Lambrama	Marjuni
Grau	Chuquibambilla	Chuqibambi-lla	Chapimarca
Abancay	Tamburco	Tamburco	Kerapata
Aymaraes	Ihuayllo	Soraya	Ihuayllo
Chincheros	Chincheros	Chincheros	Llimpe
Chincheros	Ranracancha	Ranracan-cha	Occopata
Cotabambas	Coyllurqui	Coyllurqui	Pfaco
Cotabambas	Tambomamba	Tambobam-ba	Asacasi
Acomayo	Acopia	Acopia	Huayrachapi
Quipiscanchis	Cusipata	Cusipata	Yaucat
Anta	Mollepata	Mollepata	Huamanpata
Anta	Pucyura	Pucyura	Chillaray
Calca	Pisac	Cuyogran-de	Chawattiri
Canchis	Combopata	Combopata	Chiara
Canchis	San pablo	San pablo	Alto seva songoña
Chumbivilcas	Llusco	Llusco	Marccahui
Chumbivilcas	Quiñota	Quiñota	Acoito
Paruro	Colcha	Colcha	San lorenzo
Paruro	Taray	Ccorao	Rayanniyoc
Paucartambo	Caycay	Huancarpay	Vilcabamba
Quispicanchis	Huaro	Huaro	Huaro
Arequipa	Mollepata	Mollepaya	Machaguay
Caylloma	Lluta	Huanca	Hurco
Caraveli	Yauca	Yauca	Chaviña
Caraveli	Caraveli	El molino	Huanca
La union	Pampamarca	Pampamar-ca	Jochapampa
Castilla	Machaguay	Machaguay	Cuyanca

APPENDIX-II

Access to the Telecommunications in Peru

Table-A4.5.1.1a: From where do You Use the Telephone?

		With Telephone			Without Telephone		
	TOTAL	Total	Poor	Not poor	Total	Poor	Not Poor
Community centre	37.7	38.8	36.5	40.8	35.6	37.6	32.7
Public Telephone	59.2	57.4	59.6	55.4	62.5	61.1	64.7
In House or from family/friend	1.9	2.3	2.9	1.7	1.3	0.9	2.0
Others	1.2	1.6	1.0	2.1	0.5	0.4	0.7

Table-A4.5.1.1b: Distance to the Nearest Telephone (in km)

		With Telephone			Without Telephone		
	TOTAL	Total	Poor	Not poor	Total	Poor	Not Poor
Minimum	0.0	0.0	0.0	0.0	0.0	0.0	0.5
Maximum	400.0	400.0	400.0	28.0	400.0	400.0	180.0
Mean	12.9	2.1	3.9	0.6	21.5	19.4	24.9
Std. Deviation	34.0	24.2	34.6	3.0	38.1	40.9	33.1

Table-A4.5.1.1c: Time Required to Reaching to the Nearest Public Telephone (in hours)

		With Telephone			Without Telephone		
	TOTAL	Total	Poor	Not poor	Total	Poor	Not Poor
Minimum	0	0	0	0	0	0	0
Maximum	4.500	2.750	2.750	1.033	4.500	4.500	4.500
Mean	0.305	0.018	0.018	0.018	0.587	0.565	0.629
Std. Deviation	0.894	0.147	0.187	0.087	1.184	1.144	1.260

Table-A4.5.1.1d: Transport Means to Reach to the Telephone

		With Telephone			Without Telephone		
	TOTAL	Total	Poor	Not poor	Total	Poor	Not Poor
Walking	73.5	95.5	97.1	94.0	33.4	39.7	23.8
Bicycle	4.4	1.1	1.0	1.3	10.3	13.1	6.0
Public transportation	18.3	1.1	1.0	1.3	49.5	39.7	64.2
Motorcycle	0.1	0.0	0.0	0.0	0.3	0.0	0.7
Car	0.4	0.5	0.0	0.9	0.3	0.0	0.7
Animal transportation	2.1	0.0	0.0	0.0	6.1	7.4	4.0
Others	1.3	1.8	1.0	2.6	0.3	0.0	0.7

Table-A4.5.1.1e: Alternative Means Used to Communicate if there were no Telephone

		With Telephone			Without Telephone		
	TOTAL	Total	Poor	Not poor	Total	Poor	Not Poor
Transport Agency	13.4	14.8	14.4	15.2	11.2	11.3	11.0
Letters with friends	44.8	45.2	44.7	45.7	44.2	43.1	46.2
Parcels	7.6	7.4	7.4	7.4	8.0	8.0	8.1
Radio	7.6	7.8	7.4	8.2	7.4	7.3	7.5
Travel to other village where there is a Telephone	29.6	30.0	20.2	40.3	29.0	22.3	41.6
Send mail	2.8	3.0	3.5	2.5	2.6	3.7	0.6
Own private short wave radio	0.4	0.6	0.0	1.2	0.2	0.0	0.6
Rented short wave radio	2.5	2.8	0.4	5.3	2.0	2.4	1.2
Others	4.2	3.6	4.3	2.9	5.2	7.0	1.7

Table-A4.5.1.1f: Average Expenditure You will have in the Alternative Means (in LCU)

		With Telephone			Without Telephone		
	TOTAL	Total	Poor	Not poor	Total	Poor	Not Poor
Total	7.3	6.7	5.8	7.5	8.2	9.2	6.7
Transport Agency	6.1	6.7	7.2	6.3	4.8	5.0	4.5
Letters with friends of known agency	5.0	4.3	4.5	4.0	6.1	7.2	4.1
Parcels	15.5	13.4	12.9	13.8	18.5	14.6	25.7
Communication by radio	8.4	9.9	4.4	15.2	5.9	6.0	5.9
Travel to other village where there is a Telephone	7.8	8.2	5.5	9.7	7.0	7.4	6.6
Send mail	6.2	5.8	5.6	6.2	7.0	7.2	5.0
Own private short wave radio	2.5	3.0	0.0	3.0	0.0	0.0	0.0
Rented private short wave radio	11.7	3.9	3.0	4.0	28.7	34.8	4.5
Others	14.4	7.9	5.5	11.9	21.4	22.5	13.0

Quality of Service: Peru

Table-A4.5.1.1g: How will You Rank the Telephone Service?

		With Telephone			Without Telephone		
	TOTAL	Total	Poor	Not poor	Total	Poor	Not Poor
Excellent	13.1	13.6	14.4	12.9	12.4	11.8	13.2
Good	49.6	45.7	49.8	42.1	56.6	59.4	52.3
Regular	33.7	37.6	33.5	41.2	26.6	26.6	26.5
Poor	2.7	2.9	2.4	3.4	2.4	1.7	3.3
Very Poor	0.9	0.2	0.0	0.4	2.1	0.4	4.6

Table-A4.5.1.1h: What is Your Opinion about the Telephone Service?

		With Telephone			Without Telephone		
	TOTAL	Total	Poor	Not poor	Total	Poor	Not Poor
I receive a lot more of what I pay	1.3	1.1	0.5	1.7	1.6	0.9	2.6
I receive more than what I pay	3.2	2.7	3.3	2.1	4.2	4.4	4.0
i receive what I pay, the price is fair	58.5	59.3	63.2	55.8	57.1	53.3	62.9
I receive less of what I pay	32.3	31.2	29.7	32.6	34.2	38.4	27.8
I receive a lot less of what I pay	4.7	5.7	3.3	7.7	2.9	3.1	2.6

Table-A4.5.1.1i: What are the Disadvantages of the Current Telephone Service?

		With Telephone			Without Telephone		
	TOTAL	Total	Poor	Not poor	Total	Poor	Not Poor
Too much distance	21.6	3.2	1.8	4.4	52.4	49.6	56.4
Costly	18.9	20.0	26.5	14.3	17.1	17.1	17.1
Bad communication	8.3	11.3	9.6	12.7	3.4	3.9	2.8
Don't find the person in charge	7.0	7.4	8.2	6.8	6.4	7.8	4.4
Only can call at certain hours	21.4	27.4	23.7	30.7	11.2	10.1	12.7
No disadvantage	22.8	30.6	30.1	31.1	9.6	11.6	6.6

Pricing and Affordability: Peru

Table-A4.5.1.1j: How do You Consider the Current Tariff that You Pay?

		With Telephone			Without Telephone		
	TOTAL	Total	Poor	Not poor	Total	Poor	Not Poor
Number of households	129758.5	83788.9	39709.5	44079.4	45969.6	27654.5	18315.0
Too high	42.4	39.0	37.3	40.5	48.5	48.2	49.0
Appropriate	56.2	59.4	61.2	57.8	50.4	50.9	49.7
Too low	1.4	1.6	1.4	1.7	1.1	0.9	1.3

Table-A4.5.1.1k: Transport Cost Spent to Reach the Public Telephone (in LCU)

		With Telephone			Without Telephone		
	TOTAL	Total	Poor	Not poor	Total	Poor	Not Poor
N	991	491	257	243	500	327	173
Minimum	0	0	0	0	0	0	0
Maximum	40	10	10	2	40	40	40
Mean	0.98	0.07	0.13	0.01	1.87	1.07	3.38
Std. Deviation	3.50	0.68	0.94	0.13	4.72	3.30	6.36

Table-A4.5.1.1l: Value of Travel time (in LCU)

		With Telephone			Without Telephone		
	TOTAL	Total	Poor	Not poor	Total	Poor	Not Poor
N	991	500	257	243	500	327	173
Minimum	0	0	0	0	0	0	0
Maximum	4.52	2.76	2.76	1.04	4.52	4.52	4.52
Mean	0.23	0.02	0.02	0.02	0.43	0.36	0.57
Std. Deviation	0.68	0.15	0.19	0.09	0.91	0.79	1.08

Access to Telecommunications Services: Bangladesh

Table-A4.5.1.2a: Time Required to Reach to the Nearest Public Telephone (in hours)

		With Telephone			Without Telephone		
	TOTAL	Total	Poor	Not poor	Total	Poor	Not Poor
Minimum	0	0.000	0	0	0.333	0.333	0.333
Maximum	2	0.500	0.5	0.5	2.000	2	2
Mean	0.612	0.227	0.292	0.184	0.991	0.845	1.051
Std. Deviation	0.5853	0.208	0.197	0.206	0.590	0.544	0.600

Table-A4.5.1.2b: Distance to the Nearest Public Telephone (in KM)

		With Telephone			Without Telephone		
	TOTAL	Total	Poor	Not poor	Total	Poor	Not Poor
Minimum	0	0	0	0	4	4	4
Maximum	20	2	2	2	20	20	20
Mean	4.887	1.135	1.429	0.941	8.587	7.524	9.030
Std. Deviation	5.391	0.872	0.735	0.904	5.422	5.186	5.482

Table-A4.5.1.2c: Transport Means to Reach the Telephone

		With Telephone			Without Telephone		
	TOTAL	Total	Poor	Not poor	Total	Poor	Not Poor
Walking	53.0	77.1	70.0	78.9	20.0	25.0	19.4
Public Transport	24.1				57.1	75.0	54.8
Rickshaw	20.5	20.8	30.0	18.4	20.0		22.6
Others	2.4	2.1		2.6	2.9		3.2

Table-A4.5.1.2d: Alternative Means Used to Communicate if there were no Telephone

		With Telephone			Without Telephone		
	TOTAL	Total	Poor	Not poor	Total	Poor	Not Poor
Would not communicate	2.1				4.5		4.8
Telephone from other place	20.8	19.2		22.7	22.7		23.8
Post office	16.7	26.9	25.0	27.3	4.5		4.8
Sending a person	58.3	53.8	75.0	50.0	63.6	100	61.9
Other	2.1				4.5		4.8

Table-A4.5.1.2e: Average Expenditure you will have in the Alternative Means of Communication

		With Telephone			Without Telephone		
	TOTAL	Total	Poor	Not poor	Total	Poor	Not Poor
Time in alternative means (hours)	49.5	63.5	5.5	75.1	33.5	4	34.95
Cost in alternative means (in LCU)	141.1	81.8	67.5	84.5	211.7	80	218.25

Quality of Service

Table-A4.5.1.2f: Are You Satisfied with the Current Level of Service?

	Poor	Not Poor	Total
Satisfied	42.86	45.16	44.79
Dissatisfied	42.86	19.35	23.18
Extremely Dissatisfied	14.29	35.48	32.03

Pricing and Affordability: Bangladesh

Table-A4.5.1.2g: Average Expenditure, Number of Call and Duration

		With Telephone			Without Telephone		
	TOTAL	Total	Poor	Not poor	Total	Poor	Not Poor
Avg. monthly exp. on telephone call	362.4	401.5	54.2	477.7	305.0	56.3	338.2
Avg. Number of call per month	4.8	3.6	1.8	4.0	6.6	1.8	7.2
Avg. duration per call	4.4	4.9	3.9	5.1	3.6	2.8	3.7

Table-A4.5.1.2h: Transport Cost for Telephone (in LCU)

		With Telephone			Without Telephone		
	TOTAL	Total	Poor	Not poor	Total	Poor	Not Poor
N	284	141	56	85	143	42	101
Minimum	0	0	0	0	8	8	8
Maximum	40	20	20	20	40	40	40
Mean	10.28	4.54	4.64	4.47	15.94	16.38	15.76
Std. Deviation	9.97	6.03	5.38	6.46	9.85	10.86	9.45

Table-A4.5.1.2i: Value of Travel Time (in LCU)

		With Telephone			Without Telephone		
	TOTAL	Total	Poor	Not poor	Total	Poor	Not Poor
N	284	141	56	85	143	42	101
Minimum	0.00	0.00	0.00	0.00	23.33	23.33	23.33
Maximum	120.00	45.00	45.00	45.00	120.00	120.00	120.00
Mean	41.76	17.02	21.58	14.02	66.15	58.29	69.42
Std. Deviation	35.72	16.51	15.58	16.51	32.67	32.51	32.33

Table-A4.5.2a: Summary Statistics: Peru

Variables	N	Min.	Max.	Mean	Std. Dev
Settlement has a public telephone or not	1000	0	1	0.50	0.50
Annual household expenditure in LCU	1000	302.24	98224.51	8163.70	7075.40
Distance to the public phone in KM	991	0	600	9.41	38.40
Time spent to reach the public phone (hrs).	991	0	4.5	0.31	0.89
Distance to the nearest post office in KM	991	0	200	1.77	10.22
Household size	1000	1	14	4.90	2.03
Gender of the household head (male=1, female=0)	999	0	1	0.89	0.31
Age of the household head	991	20	90	44.26	14.05
Dummy of economic activity (agriculture=1)	1000	0	1	0.65	0.48
Years of education of the household head	1000	0	17	7.49	4.63

Table-A4.5.2b: Summary Statistics: Bangladesh

Variables	N	Minimum	Maximum	Mean	Std. Dev
Settlement has a public telephone or not	284	0	1	0.50	0.50
Annual household expenditure in LCU	284	6300.00	1055000.00	81496.29	98190.44
Distance to the nearest public telephone in KM	284	0	20	4.89	5.39
Time spent to reaching the public telephone in hrs.	284	0	2	0.61	0.59
Distance to the nearest post office in KM	284	0	7	2.37	2.00
Household size	284	2	10	5.38	1.74
Gender of the household head (male=1, female=0)	284	0	1	0.94	0.23
Age of the household head	284	15	85	46.57	12.74
Dummy of economic activity (agriculture=1)	284	0	1	0.54	0.50
Education of the head of the household (illiterate=0)	284	0	1	0.84	0.37

APPENDIX-III

Table-A.6.1a: One-way Distance to the Nearest Public Telephone and Usage Rate

One Way Distance	Percent of Sample in Category	Usage Rate (%)
Bangladesh		
Less than 2 kilometres	27.1	51.95
2 to < 5 kilometres	33.5	15.79
5 to < 10 kilometres	22.5	32.81
10 kilometres and more	16.9	20.83
Total	100.0	30.28
Peru		
Less than 2 kilometres	69.64	75.00
2 to < 5 kilometres	8.50	97.62
5 to < 10 kilometres	6.58	98.46
10 kilometres and more	15.28	96.69
Total	100.00	81.78

Table-A.6.1b: One-way Transport Cost to the Nearest Public Telephone and Usage Rate

One Way Transport Cost (in LCU.)	Percent of Sample in Category	Usage Rate (%)
Bangladesh		
No transport cost	21.48	47.54
Up to Tk.10	50.70	27.08
>Tk. 10 to Tk. 20	16.90	31.25
>Tk. 20	10.92	9.68
Total	100.00	30.28
Peru		
No transport cost	80.87	78.60
>S/.0 to S/.2	8.60	95.29
>S/.2 to S/.5	5.16	92.16
>S/.5	5.36	98.11
Total	100.00	81.78

Table-A.6.3.3a: Summary Statistics: Bangladesh

	N	Minimum	Maximum	Mean	Std. Dev.
User status	284	0	1	0.303	0.460
Expenditure, yearly total, (in Tk.)	284	6300	1055000	81496.285	98190.445
Illiterate, literature, dummy	284	0	1	0.838	0.369
Occupation (ag=1, non_ag=0)	284	0	1	0.535	0.500
Age of the head of the HH	284	15	85	46.567	12.737
HH size	284	2	10	5.380	1.742
Settlement with telephone (yes=1)	284	0	1	0.496	0.501
Sex of the head of the HH	284	0	1	0.944	0.231
Valid Cases	284				

Table-A.6.3.3b: Summary Statistics: Peru

	N	Minimum	Maximum	Mean	Std. Dev.
User status	988	0	1	0.818	0.386
Settlement with telephone	1000	0	1	0.500	0.500
Age of the household head	991	20	90	44.259	14.052
Years of education of the HH head	1000	0	17	7.487	4.625
Family size	1000	1	14	4.895	2.026
House has room for econ. activity	1000	0	1	0.244	0.430
Annual household expenditure	1000	302.24	98224.51	8163.704	7075.396
Benefits of public phones:					
Fast communication (Yes=1)	996	0	1	0.611	0.488
Avoid trips (Yes=1)	996	0	1	0.217	0.412
Avoid isolation (Yes=1)	996	0	1	0.126	0.331
Money saving (Yes=1)	996	0	1	0.087	0.282
Valid Cases	988				

Table-A6.3.4a: Estimated Compensating Variation: Alternative Methods to the Service of Rural Telephony in Bangladesh[1/]

	Quartile of Expenditure			
	Bottom 25%			Top 25%
	I	II	III	IV
Alternative Methods				
	Compensated Variation per Call (Tk.)			
Telephone from other place	8.60	11.57	11.06	10.95
Sending letters (post office)	5.88	7.88	7.53	7.43
Travel	36.14	47.29	43.41	45.14
Others	66.84	89.07	80.88	87.36
Cost of Telecommunications				
	Direct + Indirect Costs per Call (Tk.)			
	55.00	30.33	42.40	49.35

[1/]Compensated Variation following Small and Rosen (1981)

Table-A6.3.4a: Estimated Compensating Variation: Alternative Methods to the Service of Rural Telephony in Peru: South Sierra[1/]

	Quartile of Expenditure			
	Bottom 25%			Top 25%
	I	II	III	IV
Alternative Methods				
	Compensated Variation per Call (S/.)			
Sending Letters (mail office)	6.81	6.52	6.13	5.83
Sending letters (transport agency)	5.48	5.29	5.10	4.98
Radio	5.65	5.35	5.13	4.91
Travel	9.82	9.27	8.52	7.00
Cost of Telecommunications				
	Direct + Indirect Costs per Call (S/.)			
	2.91	3.13	3.81	4.48

[1/]Compensated Variation following Small and Rosen (1981)

APPENDIX-IV

Table A.7.3a: Summary Statistics of Covariates: Peru

Covariates	Min.	Max	Mean	Std. Dev.
Logarithm of the tariff	0.5	2	1.236	0.5632
Logarithm of the expenditure	4.60	9.01	6.292	0.6575
Populated rural center with public phone	0	1	0.500	0.5003
Time required to arrive at the public telephone (hours)	0	4.5	0.305	0.8936
Benefits of using telephone				
Fast communication (yes=1, no=0)	0	1	0.611	0.4877
It avoids trips (yes=1, no=0)	0	1	0.217	0.4123
It avoids isolation (yes=1, no=0)	0	1	0.126	0.3315
It saves money (yes=1, no=0)	0	1	0.087	0.2825
It saves time (yes=1, no=0)	0	1	0.127	0.3326
Useful in emergency (yes=1, no=0)	0	1	0.257	0.4372
Disadvantages in the use of public telephone				
Excessive distance to the public telephone (yes=1)	0	1	0.246	0.4309
Costly (yes=1, no=0)	0	1	0.170	0.3755
Bad communication (yes=1, no=0)	0	1	0.068	0.2523
I don't find the person (yes=1, no=0)	0	1	0.063	0.2435
Timetable restriction (yes=1, no=0)	0	1	0.179	0.3833
No disadvantage (yes=1, no=0)	0	1	0.011	0.1046
Logarithm of transfers to relative (proxy of migration)	0	9.29	0.799	2.0066
Cost of alternative communication means:				
Need to travel	0	80	0.764	3.8987
Letter shipment through relatives	0	500	2.297	17.3512
Shipment charges	0	150	1.234	7.5309
Communication by radio	0	200	0.607	6.5395
Need to travel to the next populated center	0	300	2.243	10.9659
Shipment of mail	0	15	0.172	1.1983
Duration of the calls	0	4	1.630	1.1146
Educational personnel in the CCPP	0	1	0.065	0.2466
Health personnel	0	1	0.010	0.0995

Table A.7.3b: Summary Statistics of Covariates: Bangladesh

Covariates	Min.	Max.	Mean	Std. Dev.
Tariff				
3 minutes local call	1.7	18.0	9.947	6.313
1 minute NLD call	3.4	42.0	22.696	14.832
1 minute international call	26.0	100.0	63.188	28.333
Total yearly expenditure of HH (in LCU)	6300	1055000	81496.29	98190.44
Opportunity cost of Telephone (in LCU)	0	52.5	15.50	13.76
Distance of post office (in km)	0	7	2,37	2.00
Migration city	0	8	0.22	0.75
Migration abroad	0	9	0.44	1.33
sex of the head of the HH	0	1	0.94	0.23
Age of the head of the household (HH)	15	85	46.57	12.74
Age square of the head of the HH	225	7225	2330.14	1281.79

Occupation, head, dummy, non ag 1, ag	0	1	0.46	0.50
Illiterate, literature, edu, he	0	1	0.84	0.37

Table-A.7.3c: WTP for Local Calls

Country/Price		With phone			Without Phone		
Bangladesh	TOTAL	Total	Poor	Not Poor	Total	Poor	Not poor
Price 1 (Tk. 1.7)							
No	18.3	25.0	21.1	21.1	12.8	30.0	6.9
Yes	81.7	75.0	78.9	78.9	87.2	70.0	93.1
Price 2 (Tk. 7)							
No	30.4	44.4	35.0	35.0	15.2	25.0	9.5
Yes	69.6	55.6	65.0	65.0	84.8	75.0	90.5
Price 3 (Tk. 13)							
No	49.2	62.5	45.5	45.5	40.0	50.0	37.0
Yes	50.8	37.5	54.5	54.5	60.0	50.0	63.0
Price 4 (Tk. 18)							
No	66.2	79.5	77.4	77.4	48.5	50.0	47.6
Yes	33.8	20.5	22.6	22.6	51.5	50.0	52.4
Peru							
Price 1 (S/. 0.50)							
No	12.9	13.8	17.8	10.1	10.9	12.8	7.3
Yes	87.1	86.2	82.2	89.9	89.1	87.2	92.7
Price 2 (S/. 1.0)							
No	51.7	53.2	50.0	56.3	48.5	42.4	57.1
Yes	48.3	46.8	50.0	43.7	51.5	57.6	42.9
Price 3 (S/. 1.50)							
No	76.5	81.8	82.1	81.4	70.7	75.5	60.9
Yes	23.5	18.2	17.9	18.6	29.3	24.5	39.1
Price 4 (S/. 2.0)							
No	84.5	85.4	86.8	84.0	83.5	87.8	74.4
Yes	15.5	14.6	13.2	16.0	16.5	12.2	25.6

Table-A.7.3d: WTP for National Long Distance Calls

Country/Price		With phone			Without Phone		
Bangladesh	*TOTAL*	Total	Poor	Not Poor	Total	Poor	Not Poor
Price 1 (Tk. 3.4)							
No	33.8	50.0	42.1	42.1	20.5	20.0	20.7
Yes	66.2	50.0	57.9	57.9	79.5	80.0	79.3
Price 2 (Tk. 16)							
No	47.1	50.0	40.0	40.0	43.8	54.5	38.1
Yes	52.9	50.0	60.0	60.0	56.3	45.5	61.9
Price 3 (Tk. 29)							
No	81.4	100.0	100.0	100.0	68.6	50.0	74.1
Yes	18.6	0.0	0.0	0.0	31.4	50.0	25.9
Price 4 (Tk. 42)							
No	88.2	100.0	100.0		71.9	90.9	61.9
Yes	11.8	0.0	0.0		28.1	9.1	38.1

Peru							
Price 1 (S/. 1.0)							
No	25.7	26.8	23.9	29.5	23.3	26.7	17.1
Yes	74.3	73.2	76.1	70.5	76.7	73.3	82.9
Price 2 (S/. 1.50)							
No	57.9	58.5	57.7	59.2	56.7	51.6	64.3
Yes	42.1	41.5	42.3	40.8	43.3	48.4	35.7
Price 3 (S/. 2.0)							
No	80.7	85.9	91.1	79.1	75.0	77.7	69.6
Yes	19.3	14.1	8.9	20.9	25.0	22.3	30.4
Price 4 (S/. 3.0)							
No	86.4	88.3	86.5	90.2	84.1	86.7	78.6
Yes	13.6	11.7	13.5	9.8	15.9	13.3	21.4

Table-A.7.3e: WTP for International Calls

Country/Price		With phone			Without Phone		
Bangladesh	TOTAL	Total	Poor	Not Poor	Total	Poor	Not Poor
Price 1 (Tk. 26)							
No	24.3	34.4	26.3	26.3	15.8	22.2	13.8
Yes	75.7	65.6	73.7	73.7	84.2	77.8	86.2
Price 2 (Tk. 51)							
No	36.8	36.1	40.0	40.0	37.5	63.6	23.8
Yes	63.2	63.9	60.0	60.0	62.5	36.4	76.2
Price 3 (Tk. 75)							
No	66.1	79.2	72.7	72.7	57.1	50.0	59.3
Yes	33.9	20.8	27.3	27.7	42.9	50.0	40.7
Price 4 (Tk. 100)							
No	66.2	70.5	61.3	61.3	60.6	75.0	52.4
Yes	33.8	29.5	38.7	38.7	39.4	25.0	47.6

Table-A.7.4.1a: WTP for the Local Rural Public Telephone Service in Bangladesh, Logit Model, *Dependent Variable: Accept to pay bj Taka (Standard Errors are in the parenthesis)*

Explanatory Variables	Model 1	Model 2	Model 3
Tariff	-0.23340	-0.23920	-0.24200
	(0.0278)**	(0.0286)**	(0.0291)**
Logarithm of HH expenditure	0.14080	0.11310	-0.03430
	(0.1917)	(0.1944)	(0.20560)
Opportunity cost of tel use (travel time and travel exp for tel)	0.33210 (0.1313)*	0.41360 (0.1446)**	0.42320 (0.1493)**
Distance of post office (alternative communication means)		-0.10510 (0.0828)	-0.14370 (0.08720)
Family members migrated to city (yes=1)		0.17350	0.23410
		(0.2449)	(0.26570)
Family members migrated abroad (yes=1)		0.15320	0.15340
		(0.1247)	(0.13100)
Household characteristics			
Gender of the head of the HH (male=1)			0.14880
			(0.66470)
Age of the head of the household			0.06270

			(0.07990)
Age square of the head of the household			-0.00070
			(0.00080)
Occupation of the head of the household			0.25210
			(0.32720)
Education of the head of the household			1.07040
			(0.4556)*
Constant	0.55660	0.89340	0.05000
	(2.1416)	(2.1498)	(2.95050)
Log likelihood	-137.128	-134.991	-130.62
Number of observations	282	282	282
Wald chi2	108.47	112.74	121.48
Probability>chi2	0.000	0.000	0.000

***, **, *: Significant at 1%, 5% and 10% respectively.

Table-A.7.4.1b: WTP for the Local Rural Public Telephone Service in Bangladesh, Log Logit Model, *Dependent Variable: Accept to pay bj Taka (Standard Errors are in the parenthesis)*

Explanatory Variables	Model 1	Model 3	Model 5
Logarithm of Tariff	-2.03620	-2.08060	-2.10080
	(0.2797)**	(0.2875)**	(0.2908)**
Logarithm of HH expenditure	0.12830	0.11020	-0.02490
	(0.1931)	(0.1951)	(0.2057)
Opportunity cost of tel use (travel time and travel exp for tel)	0.36230	0.43830	0.46190
	(0.1348)**	(0.1472)**	(0.1518)**
Distance of post office (alternative communication means)		-0.10430	-0.14570
		(0.0827)	(0.087)
Family members migrated to city (yes=1)		0.15010	0.2062
		(0.2264)	(0.2445)
Family members migrated abroad (yes=1)		0.12820	0.1301
		(0.1262)	(0.1331)
Household characteristics			
Gender of the head of the HH (male=1)			0.1468
			(0.6832)
Age of the head of the household			0.08080
			(0.082)
Age square of the head of the household			-0.00080
			(0.0008)
Occupation of the head of the household			0.20090
			(0.3267)
Education of the head of the household			1.06230
			(0.4683)*
Constant	2.59200	2.88830	1.49310
	(2.212)	(2.2183)	(3.0367)
Log likelihood	-133.131	-131.305	-127.060
Number of observation	282	282	282
Probability>chi2	0.000	0.000	0.000

***, **, *: Significant at 1%, 5% and 10% respectively.

Table-A.7.4.1c: WTP for the Local Rural Public Telephone Services in Bangladesh: Bounded Probit Model

Dependent Variable: Accept to pay bj taka, (Standard errors in the parenthesis)

Explanatory Variables	Model 1	Model 3	Model 5
Tariff	-1.12780	-1.14870	-1.16850
	(0.1388)**	(0.1418)**	(0.1455)**
Logarithm of HH expenditure	0.09090	0.07550	-0.00590
	(0.1138)	(0.1161)	(0.1223)
Opportunity cost of telephone use (travel time and travel exp for tel)	0.20820	0.25240	0.27740
	(0.0779)**	(0.0848)**	(0.0888)**
Distance of post office (alternative communication means)		-0.05810	-0.08860
		(0.0483)	(0.0516)
Family members migrated to city (yes=1)		0.10420	0.13130
		(0.1375)	(0.1466)
Family members migrated abroad (yes=1)		0.07630	0.07630
		(0.072)	(0.0741)
Household characteristics			
Gender of the head of the household (male=1)			0.0978
			(0.3887)
Age of the head of the household			0.05480
			(0.0478)
Age square of the head of the household			-0.00050
			(0.0005)
Occupation of the head of the household			0.08500
			(0.1929)
Education of the head of the household			0.65040
			(0.2644)*
Constant	-5.6283	-5.595	-6.91410
	(1.1713)**	(1.1853)**	(1.7515)**
Log likelihood	-133.492	-131.661	-127.172
Number of observations	282	282	282
Sigma	-0.887	-0.871	-0.856
Wald chi2	115.74	119.40	128.38
Probability>chi2	0.000	0.000	0.000

***, **, *: Significant at 1%, 5% and 10% respectively.

Table-A.7.4.1d: WTP for the Local Rural Public Telephone Service in Peru, Logit Model

Dependent Variable: Accept to pay bj soles (Standard Errors are in the parenthesis)

Explanatory Variables	Model 1	Model 2	Model 3
Logarithm of the tariff	-2.77739***	-2.84575***	-2.87252***
	(0.1973)	(0.2040)	(0.2015)
Logarithm of the expenditure	0.29779**	0.30216**	0.28087**
	(0.1303)	(0.1361)	(0.1411)
Populated rural center with public phone	-0.39935**	-0.43023**	-0.39988**
	(0.1991)	(0.2029)	(0.2102)
Time required to arrive at the public telephone (hours)	0.45651**	0.46932***	0.46758***
	(0.1006)	(0.1024)	(0.1027)

Benefits of using telephone			
Fast communication (yes=1, no=0)	0.55333***	0.59184***	0.59676***
	(0.2010)	(0.2003)	(0.2035)
It avoids trips (yes=1, no=0)	0.88759***	0.94684***	0.96830***
	(0.2350)	(0.2341)	(0.2363)
It avoids isolation (yes=1, no=0)	0.15696	0.19130	0.15721
	(0.2635)	(0.2694)	(0.2790)
It saves money (yes=1, no=0)	0.15737	0.11321	0.20279
	(0.3133)	(0.3185)	(0.3232)
It saves time (yes=1, no=0)	-0.15466	-0.19121	-0.19039
	(0.2457)	(0.2432)	(0.2458)
Useful in emergency (yes=1, no=0)	0.45609**	0.46382**	0.47100**
	(0.2245)	(0.2278)	(0.2304)
Disadvantages in the use of public telephone			
Excessive distance to the public telephone (yes=1, no=0)	-0.62998***	-0.62773***	-0.52471**
	(0.2320)	(0.2341)	(0.2502)
Costly (yes=1, no=0)	-0.00374	0.01668	0.00483
	(0.2259)	(0.2296)	(0.2389)
Bad communication (yes=1, no=0)	-0.31496	-0.35624	-0.35053
	(0.3424)	(0.3401)	(0.3430)
I don't find the person (yes=1, no=0)	0.83522**	0.80963**	0.73633**
	(0.3480)	(0.3444)	(0.3561)
Timetable restriction (yes=1, no=0)	-0.31154	-0.25722	-0.29753
	(0.2226)	(0.2247)	(0.2254)
No disadvantage (yes=1, no=0)	0.99830*	0.94524	1.03079*
	(0.5889)	(0.6046)	(0.6055)
Logarithm of transfers to relative (proxy of migration)		0.04069	0.03610
		(0.0423)	(0.0442)
Cost of alternative communication means:			
Need to travel		0.03915**	0.03602*
		(0.0198)	(0.0203)
Letter shipment through relatives		-0.00112	-0.00089
		(0.0036)	(0.0035)
Shipment charges		0.00055	0.00208
		(0.0134)	(0.0131)
Communication by radio		-0.06093	-0.05199
		(0.0407)	(0.0402)
Need to travel to the next populated center		0.00715	0.00775*
		(0.0046)	(0.0046)
Shipment of mail		0.09393	0.07445
		(0.0720)	(0.0805)
Duration of the calls			-0.08575
			(0.0796)
Educational personnel in the CCPP			1.17880***
			(0.4100)
Health personnel			-0.98858
			(0.6379)
Constant	0.84683	0.79902	1.00737
	(0.8109)	(0.8508)	(0.8785)
Log likelihood	-469.96	-463.92	-456.09

Number of observations	971	971	971
Wald chi2	215.2	214.4	233.52
Probability > chi2	0.000	0.000	0.000

***, **, *: Significant at 1%, 5% and 10% respectively.

Table-A.7.4.1e: WTP for the Local Rural Public Telephone Service in Peru, Bounded Probit Model, *Dependent Variable: Accept to pay bj soles (Standard Errors are in the parenthesis)*

Explanatory Variables	Model 1	Model 2	Model 3
Logarithm of the upperbound tariff	-1.7639 ***	-1.79308***	-1.80866***
	(0.1067	(0.1091)	(0.1088)
Logarithm of the expenditure	0.1841 **	0.18725**	0.17203**
	(0.0759	(0.0789)	(0.0814)
Populated rural center with public phone	-0.2335 **	-0.24972**	-0.23986**
	(0.1170	(0.1187)	(0.1224)
Time required to arrive at the public telephone (hours)	0.2668 ***	0.27370***	0.26912***
	(0.0568	(0.0575)	(0.0578)
Benefits of using telephone			
Fast communication (yes=1, no=0)	0.3381 ***	0.36251***	0.36376***
	(0.1186	(0.1186)	(0.1202)
It avoids trips (yes=1, no=0)	0.5171 ***	0.54737***	0.55452***
	(0.1364	(0.1365)	(0.1371)
It avoids isolation (yes=1, no=0)	0.0934	0.11330	0.10651
	(0.1526	(0.1575)	(0.1613)
It saves money (yes=1, no=0)	0.1293	0.10480	0.15309
	(0.1799	(0.1818)	(0.1832)
It saves time (yes=1, no=0)	-0.0626	-0.08399	-0.08123
	(0.1452	(0.1444)	(0.1460)
Useful in emergency (yes=1, no=0)	0.2659 **	0.26543**	0.27636**
	(0.1278	(0.1286)	(0.1293)
Disadvantages in the use of public telephone			
Excessive distance to the public telephone (yes=1, no=0)	-0.3669 ***	-0.36201***	-0.30923**
	(0.1363	(0.1372)	(0.1461)
Costly (yes=1, no=0)	0.0019	0.01189	0.00277
	(0.1330	(0.1344)	(0.1377)
Bad communication (yes=1, no=0)	-0.1563	-0.18052	-0.19373
	(0.2003	(0.1974)	(0.1966)
I don't find the person (yes=1, no=0)	0.4192 **	0.39818**	0.36852*
	(0.2021	(0.2033)	(0.2050)
Timetable restriction (yes=1, no=0)	-0.1446	-0.11362	-0.14855
	(0.1306	(0.1311)	(0.1314)
No disadvantage (yes=1, no=0)	0.6315 *	0.59912	0.65028*
	(0.3775	(0.3880)	(0.3887)
Log of transfers to relative (proxy of migration)		0.02367	0.02045
		(0.0247)	(0.0256)
Cost of alternative communication means:			
Need to travel		0.01982	0.01817
		(0.0133)	(0.0133)
Letter shipment through		-0.00038	-0.00030

relatives			
		(0.0024)	(0.0024)
Shipment charges		0.00039	0.00119
		(0.0075)	(0.0074)
Communication by radio		-0.03525	-0.02995
		(0.0245)	(0.0249)
Need to travel to the next populated center		0.00388	0.00398
		(0.0032)	(0.0031)
Shipment of mail		0.05968	0.04941
		(0.0382)	(0.0411)
Duration of the calls			-0.03977
			(0.0468)
Educational personnel in the CCPP			0.67465***
			(0.2112)
Health personnel			-0.55601
			(0.3955)
Constant	-9.4883 ***	-9.69683***	-9.65643***
	(0.6926	(0.7118)	(0.7215)
Log likelihood	-459.0	-453.64	-446.17
Number of observations	97	971	971
Sigma	-0.56	-0.558	-0.553
Wald chi2	295.	294.1	314.28
Probability > chi2	0.00	0.000	0.000

***, **, *: Significant at 1%, 5% and 10% respectively.

Table-A.7.4.1f: WTP for the NLD rural public telephone service in Bangladesh, Logit Model, *Dependent Variable: Accept to pay bj Taka (Standard Errors are in the parenthesis)*

Explanatory Variables	Model 1	Model 2	Model 3
Tariff	-0.13180	-0.13520	-0.13680
	(0.0150)**	(0.0155)**	(0.0159)**
Logarithm of HH expenditure	0.49170	0.47780	0.39330
	(0.2130)*	(0.2210)*	(0.23040)
Opportunity cost of tel use (travel time and travel exp for tel)	0.36820	0.25480	0.26720
	(0.1449)*	(0.15700)	(0.16340)
Distance of post office (alternative communication means)		0.26080	0.22210
		(0.0918)**	(0.0955)*
Family members migrated to city (yes=1)		0.10880	0.12100
		(0.19640)	(0.18730)
Family members migrated abroad (yes=1)		0.11180	0.10680
		(0.15730)	(0.16420)
Household characteristics			
Gender of the head (male=1)			0.62950
			(0.74960)
Age of the head of the household			0.03790
			(0.08950)
Age square of the head			-0.00040
			(0.0009)
Occupation of the head			-0.17730

			-0.36480
Education of the head			1.03310
			(0.5236)*
Constant	-3.75690	-3.97270	-5.23830
	(2.33000)	(2.40440)	(3.20550)
Log likelihood	-116.444	-112.048	-109.458
Number of observations	280	280	280
Wald chi2	150.10	158.89	164.07
Probability>chi2	0.000	0.000	0.000

***, **, *: Significant at 1%, 5% and 10% respectively.

Table-A.7.4.1g: WTP for the NLD Rural Public Telephone Service in Bangladesh, Log Logit Model, *Dependent Variable: Accept to pay bj Taka (Standard Errors are in the parenthesis)*

Explanatory Variables	Model 1	Model 2	Model 3
Logarithm of Tariff	-2.25920	-2.32590	-2.41750
	(0.2749)**	(0.2825)**	(0.2969)**
Logarithm of HH expenditure	0.48850	0.48440	0.39010
	(0.2187)*	(0.2281)*	(0.2373)~
Opportunity cost of tel use (travel time and travel exp for tel)	0.43690	0.30530	0.34740
	(0.1601)**	(0.1746)	(0.1804)
Distance of post office (alternative communication means)		0.28570	0.23930
		(0.0940)**	(0.0984)*
Family members migrated to city (yes=1)		0.12280	0.13810
		(0.1951)	(0.1897)
Family members migrated abroad (yes=1)		0.09990	0.10150
		(0.1601)	(0.1695)
Household characteristics			
Gender of the head (male=1)			0.729
			-0.84010
Age of the head of the household			0.09550
			-0.10120
Age square of the head			-0.00100
			-0.00100
Occupation of the head			-0.28270
			-0.38500
Education of the head			1.27230
			(0.6116)*
Constant	-0.2501	-0.48680	-3.05230
	(2.4331)	(2.4958)	(3.4632)
Log likelihood	-107.975	-103.034	-99.627
Number of observations	280	280	280
Wald chi2	167.04	176.92	183.73
Probability>chi2	0.000	0.000	0.000

***, **, *: Significant at 1%, 5% and 10% respectively.

Table-A.7.4.1h: WTP for the NLD Rural Public Telephone Services in Bangladesh: Bounded Probit, *Dependent Variable: Accept to pay bj Taka (Standard Errors are in the parenthesis)*

Explanatory Variables	Model 1	Model 3	Model 5
Tariff	-1.25800	-1.29560	-1.33900
	(0.1321)**	(0.1363)**	(0.1441)**
Logarithm of HH expenditure	0.29690	0.28240	0.23010
	(0.1262)*	(0.1298)*	(0.1356)~
Opportunity cost of tel use (travel time and travel exp for tel)	0.26340 (0.0890)**	0.18440 (0.0963)~	0.21440 (0.1003)*
Distance of post office (alternative communication means)		0.16430 (0.0528)**	0.13610 (0.0553)*
Family members migrated to city (yes=1)		0.07030	0.07850
		(0.1205)	(0.1146)
Family members migrated abroad (yes=1)		0.06050	0.06330
		(0.0896)	(0.094)
Household characteristics			
Gender of the head (male=1)			0.4098
			(0.4857)
Age of the head			0.06390
			(0.0556)
Age square of the head			-0.00060
			(0.0006)
Occupation of the head			-0.18630
			(0.2191)
Education of the head			0.69130
			(0.3430)*
Constant	-7.5813	-7.8527	-9.94550
	(1.2632)**	(1.3064)**	(1.9754)**
Log likelihood	-107.367	-102.316	-98.942
Number of observations	280	280	280
Sigma	-0.795	-0.772	-0.747
Wald chi2	168.26	178.36	185.11
Probability>chi2	0.000	0.000	0.000

***, **, *: Significant at 1%, 5% and 10% respectively.

Table-A.7.4.1i: WTP for the LDN Rural Public Telephone Service in Peru, Logit Model *Dependent Variable: Accept to pay bj soles (Standard Errors are in the parenthesis)*

Explanatory Variables	Model 1	Model 2	Model 3
Logarithm of the tariff	-2.29331***	-2.35617***	-2.35914***
	(0.1773)	(0.1812)	(0.1819)
Logarithm of the expenditure	0.27618**	0.27868**	0.25008**
	(0.1271)	(0.1304)	(0.1343)
Populated rural center with public phone	-0.54657***	-0.60712***	-0.62334***
	(0.1936)	(0.1973)	(0.2005)
Time required to arrive at the public telephone (hours)	0.20559** (0.1020)	0.20949** (0.1027)	0.20301** (0.1026)
Benefits of using telephone			
Fast communication (yes=1)	0.04050	0.06846	0.06729

	(0.1896)	(0.1898)	(0.1907)
It avoids trips (yes=1, no=0)	0.64158***	0.66937***	0.67554***
	(0.2174)	(0.2150)	(0.2156)
It avoids isolation (yes=1, no=0)	0.52548**	0.50563**	0.50744**
	(0.2395)	(0.2466)	(0.2461)
It saves money (yes=1, no=0)	0.19004	0.12491	0.15526
	(0.2989)	(0.2989)	(0.2998)
It saves time (yes=1, no=0)	-0.25531	-0.29455	-0.27863
	(0.2341)	(0.2321)	(0.2302)
Useful in emergency (yes=1)	0.05509	0.07725	0.09451
	(0.1975)	(0.1976)	(0.1973)
Disadvantages in the use of public telephone			
Excessive distance (yes=1, no=0)	-0.59307***	-0.57369***	-0.59135**
	(0.2381)	(0.2422)	(0.2563)
Costly (yes=1, no=0)	0.29419	0.34690	0.32261
	(0.2279)	(0.2321)	(0.2378)
Bad communication (yes=1, no=0)	0.19252	0.19117	0.16813
	(0.3240)	(0.3286)	(0.3295)
I don't find the person (yes=1)	0.78628**	0.78335***	0.75383**
	(0.3044)	(0.3012)	(0.3028)
Timetable restriction (yes=1)	0.13137	0.21191	0.17618
	(0.2168)	(0.2216)	(0.2259)
No disadvantage (yes=1, no=0)	0.88434	0.85975	0.87610
	(0.8419)	(0.8339)	(0.8366)
Logarithm of transfers to relative (proxy of migration)		0.04497	0.04173
		(0.0403)	(0.0403)
Cost of alternative communication means:			
Need to travel		0.05352**	0.05259**
		(0.0218)	(0.0213)
Letter shipment through relatives		-0.00868	-0.00858
		(0.0077)	(0.0077)
Shipment charges		-0.01115	-0.01071
		(0.0107)	(0.0107)
Communication by radio		-0.03292	-0.03079
		(0.0249)	(0.0217)
Need to travel to the next populated center		0.00692	0.00667
		(0.0052)	(0.0051)
Shipment of mail		0.10283	0.09268
		(0.0690)	(0.0733)
Duration of the calls			0.03197
			(0.0779)
Educational personnel in the CCPP			0.33650
			(0.3729)
Health personnel			-0.27426
			(0.6157)
Constant	1.55849*	1.56303*	1.69737**
	(0.8010)	(0.8159)	(0.8313)
Log likelihood	-500.21	-491.92	-491.11
Number of observations	968	968	968

Wald chi2	186.4	194.7	199.1
Probability > chi2	0.000	0.000	0.000

***, **, *: Significant at 1%, 5% and 10% respectively.

Table-A.7.4.1j: WTP for the LDN Rural Public Telephone Service in Peru, Bounded Probit Model, *Dependent Variable: Accept to pay bj soles (Standard Errors are in the parenthesis)*

Explanatory Variables	Model 1	Model 2	Model 3
Logarithm of the upperbound tariff	-2.13046***	-2.17636***	-2.17996***
	(0.1451)	(0.1476)	(0.1480)
Logarithm of the expenditure	0.16948**	0.17024**	0.15139**
	(0.0746)	(0.0771)	(0.0792)
Populated rural center with public phone	-0.32481***	-0.36003***	-0.37539***
	(0.1132)	(0.1148)	(0.1168)
Time required to arrive at the public phone (hours)	0.12379**	0.12552**	0.12055**
	(0.0570)	(0.0574)	(0.0576)
Benefits of using telephone			
Fast communication (yes=1)	0.02174	0.04004	0.03821
	(0.1117)	(0.1116)	(0.1120)
It avoids trips (yes=1, no=0)	0.37917***	0.39248***	0.39674***
	(0.1271)	(0.1262)	(0.1264)
It avoids isolation (yes=1, no=0)	0.30369**	0.29012**	0.29267**
	(0.1411)	(0.1462)	(0.1458)
It saves money (yes=1, no=0)	0.11960	0.08024	0.09725
	(0.1741)	(0.1746)	(0.1745)
It saves time (yes=1, no=0)	-0.16016	-0.18856	-0.18026
	(0.1389)	(0.1377)	(0.1368)
Useful in emergency (yes=1)	0.03791	0.04410	0.05710
	(0.1168)	(0.1157)	(0.1155)
Disadvantages in the use of public telephone			
Excessive distance (yes=1)	-0.35072**	-0.33734**	-0.35077**
	(0.1384)	(0.1400)	(0.1476)
Costly (yes=1, no=0)	0.18404	0.21314	0.19810
	(0.1316)	(0.1331)	(0.1357)
Bad communication (yes=1)	0.12734	0.12258	0.10604
	(0.1914)	(0.1922)	(0.1926)
I don't find the person (yes=1,)	0.44999**	0.44057**	0.42831**
	(0.1784)	(0.1775)	(0.1779)
Timetable restriction (yes=1)	0.09331	0.13978	0.11533
	(0.1283)	(0.1303)	(0.1324)
No disadvantage (yes=1, no=0)	0.47024	0.45359	0.46863
	(0.4814)	(0.4811)	(0.4826)
Log of transfers to relative (proxy of migration)		0.02856	0.02542
		(0.0238)	(0.0238)
Cost of alternative communication means:			
Need to travel		0.03023**	0.02979**
		(0.0136)	(0.0134)
Letter shipment through relatives		-0.00454	-0.00454

		(0.0040)	(0.0040)
Shipment charges		-0.00747	-0.00714
		(0.0059)	(0.0059)
Communication by radio		-0.01839	-0.01718
		(0.0150)	(0.0122)
Need to travel to the next populated center		0.00395	0.00373
		(0.0036)	(0.0035)
Shipment of mail		0.06454*	0.05925
		(0.0377)	(0.0392)
Duration of the calls			0.02397
			(0.0459)
Educational personnel in the CCPP			0.20417
			(0.2084)
Health personnel			-0.11147
			(0.3883)
Constant	-10.17093***	-10.41314***	-10.34388***
	(0.7668)	(0.7880)	(0.7945)
Log likelihood	-497.21	-489.15	-488.25
Number of observations	968	968	968
Sigma	-0.469	-0.459	-0.459
Wald chi2	240.9	251.6	256.8
Probability > chi2	0.000	0.000	0.000

***, **, *: Significant at 1%, 5% and 10% respectively.

Table-A.7.4.1k: WTP for the International Rural Public Telephone Services in Bangladesh: Logit Model, *Dependent Variable: Accept to pay bj taka (Standard errors in the parenthesis)*

Explanatory Variables	Model 1	Model 2	Model 3
Tariff	-0.03350	-0.03710	-0.037
	(0.0052)**	(0.0056)**	(0.0057)**
Logarithm of HH expenditure	0.76860	0.65310	0.5703
	(0.1919)**	(0.1993)**	(0.2054)**
Opportunity cost of tel use (travel time and travel exp)	0.27190	0.37430	0.3831
	(0.1181)*	(0.1376)**	(0.1422)**
Distance of post office (alternative communication)		0.11720	0.1089
		(0.07710)	(0.08050)
Family members migrated to city (yes=1)		-0.01870	-0.0068
		(0.17920)	(0.17990)
Family members migrated abroad (yes=1)		0.75040	0.75
		(0.2124)**	(0.2112)**
Household characteristics			
Gender of the head (male=1)			0.1099
			(0.64560)
Age of the head of the household			0.1113
			(0.07720)
Age square of the head			-0.001
			(0.00080)
Occupation of the head			0.3171
			(0.31020)

Education of the head			0.4914
			(0.44050)
Constant	-6.79270	-6.07380	-8.6947
	(2.1066)**	(2.1750)**	(2.9379)**
Log likelihood	-158.269	-146.079	-143.381
Number of observations	280	280	280
Wald chi2	71.27	95.65	101.04
Probability>chi2	0.000	0.000	0.000

***, **, *: Significant at 1%, 5% and 10% respectively.

Table-A.7.4.11: WTP for the International Rural Public Telephone Services in Bangladesh: Log Logit Model, *Dependent Variable: Accept to pay bj taka (Standard errors in the parenthesis)*

Explanatory Variables	Model 1	Model 2	Model 3
Logarithm of Tariff	-1.96840	-2.15040	-2.1547
	(0.2994)**	(0.3201)**	(0.3248)**
Logarithm of HH expenditure	0.77140	0.66060	0.5843
	(0.1927)**	(0.2002)**	(0.2061)**
Opportunity cost of tel use (travel time and travel exp)	0.28450	0.39080	0.4051
	(0.1192)*	(0.1396)**	(0.1441)**
Distance of post office (alternative communication)		0.11950	0.109
		(0.0775)	(0.0811)
Family members migrated to city (yes=1)		-0.01130	0.0026
		(0.1792)	(0.1801)
Family members migrated abroad (yes=1)		0.75610	0.7561
		(0.2144)**	(0.2137)**
Household characteristics			
Gender of the head (male=1)			0.0557
			(0.644)
Age of the head of the household			0.1253
			(0.0794)
Age square of the head			-0.0011
			(0.0008)
Occupation of the head			0.3099
			(0.3138)
Education of the head			0.4757
			(0.4462)
Constant	-1.0127	0.14960	-2.7713
	(2.2761)	(2.355)	(3.0369)
Log likelihood	-156.079	-144.076	-141.313
Number of observations	280	280	280
Wald chi2	75.65	99.65	105.18
Probability>chi2	0.000	0.000	0.000

***, **, *: Significant at 1%, 5% and 10% respectively.

Table-A.7.4.1m: WTP for the International Rural Public Telephone Services in Bangladesh: Bounded Probit Model, *Dependent Variable: Accept to pay bj taka (Standard errors are in the parenthesis)*

Explanatory Variables	Model 1	Model 2	Model 3
Tariff	-1.04820	-1.14390	-1.1472
	(0.1519)**	(0.1604)**	(0.1631)**
Logarithm of HH expenditure	0.44570	0.37010	0.3199
	(0.1090)**	(0.1121)**	(0.1160)**
Opportunity cost of tel use (travel time and	0.16900	0.22550	0.2346
travel exp)	(0.0696)*	(0.0803)**	(0.0832)**
Distance of post office (alternative		0.06730	0.0632
communication)		(0.0454)	(0.0477)
Family members migrated to city (yes=1)		-0.00870	-0.0004
		(0.1122)	(0.1111)
Family members migrated abroad (yes=1)		0.41200	0.4083
		(0.1081)**	(0.1073)**
Household characteristics			
Gender of the head (male=1)			0.0313
			(0.3785)
Age of the head of the household			0.0735
			(0.0467)
Age square of the head			-0.0007
			(0.0005)
Occupation of the head			0.1632
			(0.1843)
Education of the head			0.3121
			(0.2656)
Constant	-6.278	-6.2782	-8.1242
	(1.0309)**	(1.0605)**	(1.6212)**
Log likelihood	-156.444	-144.664	-141.819
Number of observations	280	280	280
Sigma	-0.954	-0.874	-0.872
Wald chi2	74.92	98.48	104.17
Probability>chi2	0.000	0.000	0.000

***, **, *: Significant at 1%, 5% and 10% respectively.

PARAMETRIC ESTIMATION:

Table-A7.4.2a: Estimated WTP for the Local Telephone Service in Bangladesh, Logit

Model 1		WTP	Std Error	Confidence Interval	
Mean WTP	-∝<WTP<+∝	12.18766	0.67206	10.83711	13.53821
Truncated Mean WTP	0<WTP<+∝	12.42984	0.65098	11.12164	13.73803
Truncated Mean WTP	0<WTP< Bmax	9.884354	0.29360	9.29434	10.47437
Model 2					
Mean WTP	-∝<WTP<+∝	12.20645	0.636896	10.92656	13.48634
Truncated Mean WTP	0<WTP<+∝	12.42613	0.6091145	11.20207	13.65019
Truncated Mean WTP	0<WTP< Bmax	9.93987	0.2976509	9.341722	10.53803
Model 3					
Mean WTP	-∝<WTP<+∝	12.26135	0.6255398	11.00428	13.51842
Truncated Mean WTP	0<WTP<+∝	12.46848	0.5988251	11.26509	13.67186
Truncated Mean WTP	0<WTP< Bmax	9.980245	0.3467345	9.283456	10.67703

Table- A7.4.2b: Estimated WTP for the Local Telephone Service in Bangladesh, Log Logit

		WTP	Std. Error	Confidence Interval	
Model 1					
Truncated Median, Log Transformed		10.61779	0.78997	9.03030	12.20529
Truncated Mean, Log Transformed	-∝<lnWTP<+∝	16.38832	3.50347	9.34785	23.42881
Model 2					
Truncated Median, Log Transformed		10.65910	0.76359	9.12460	12.19359
Truncated Mean, Log Transformed	-∝<lnWTP<+∝	16.12476	2.79042	10.51720	21.73233
Model 3					
Truncated Median, Log Transformed		10.67498	0.74961	9.16859	12.18137
Truncated Mean, Log Transformed	-∝<lnWTP<+∝	16.00910	2.55435	10.87594	21.14227

Table-A7.4.2c: Estimated WTP for the Local Telephone Service in Bangladesh: Bounded Probit Model

	WTP	Std. Error	Confidence Interval	
Model 1				
Median WTP	10.47996	0.75619	8.96033	11.99958
Mean WTP	14.46335	1.39882	11.65232	17.27437
Model 2				
Median WTP	10.51213	0.76360	8.97762	12.04663
Mean WTP	14.34846	1.43632	11.46207	17.23486
Model 3				
Median WTP	10.58555	0.79099	8.99600	12.17510
Mean WTP	14.30359	1.40691	11.47631	17.13088

Table-A7.4.2d: Estimated WTP for the Local Telephone Service in Peru: Logit Model

		WTP	Std. Error	Confidence Interval (95%)	
Model 1					
Mean WTP	-∝<WTP<+∝	1.138	0.0300	1.0792	1.1970
Truncated Mean WTP	0<WTP<+∝	1.153	0.0288	1.0966	1.2095
Truncated Mean WTP	0<WTP< Bmax	0.844	0.0121	0.8203	0.8679
Model 2					
Mean WTP	-∝<WTP<+∝	1.135	0.0299	1.0768	1.1938
Truncated Mean WTP	0<WTP<+∝	1.149	0.0287	1.0926	1.2052
Truncated Mean WTP	0<WTP< Bmax	0.848	0.0125	0.8239	0.8729
Model 3					
Mean WTP	-∝<WTP<+∝	1.139	0.0299	1.0804	1.1975
Truncated Mean WTP	0<WTP<+∝	1.152	0.0288	1.0955	1.2083
Truncated Mean WTP	0<WTP< Bmax	0.850	0.0126	0.8257	0.8752

Table-A7.4.2e: Estimated WTP for the Local Rural Telephone Service in Peru: Bounded Probit Model

	WTP	Std. Error	Confidence Interval (95%)	
Model 1				
Median WTP	1.014	0.0280	0.9587	1.0686
Mean WTP	1.192	0.0763	1.0161	1.3676
Model 2				
Median WTP	1.011	0.0281	0.9558	1.0659
Mean WTP	1.182	0.0738	1.0061	1.3576
Model 3				
Median WTP	1.013	0.0281	0.9576	1.0677
Mean WTP	1.182	0.0732	1.0065	1.3580

Table-A.7.4.2f: Estimated WTP for the NLD Rural Telephone Service in Bangladesh, Logit Model

		WTP	Std. Error	Confidence Interval	
Model 1					
Mean WTP	$-\infty<WTP<+\infty$	18.59947	1.285633	16.0159	21.18305
Truncated Mean WTP	$0<WTP<+\infty$	19.2265	1.207225	16.80049	21.65251
Truncated Mean WTP	$0<WTP<$ Bmax	18.38523	0.9994806	16.37669	20.39376
Model 2					
Mean WTP	$-\infty<WTP<+\infty$	18.62171	1.343655	15.92153	21.32188
Truncated Mean WTP	$0<WTP<+\infty$	19.19478	1.260935	16.66084	21.72873
Truncated Mean WTP	$0<WTP<$ Bmax	18.41486	1.066509	16.27163	20.55809
Model 3					
Mean WTP	$-\infty<WTP<+\infty$	18.51995	1.419059	15.66824	21.37165
Truncated Mean WTP	$0<WTP<+\infty$	19.0787	1.313443	16.43924	21.71817
Truncated Mean WTP	$0<WTP<$ Bmax	18.33963	1.16563	15.99721	20.68205

Table-A.7.4.2g: Estimated WTP for the NLD Rural Telephone Service in Bangladesh, Log Logit Model

		WTP	Std. Error	Confidence Interval	
Model 1					
Truncated Median, Log Transformed		14.74	1.32830	12.068	17.407
Truncated Mean, Log Transformed	$-\infty<\ln WTP<+\infty$	20.83	1.69317	17.428	24.233
Model 3					
Truncated Median, Log Transformed		14.67	1.32663	12.000	17.332
Truncated Mean, Log Transformed	$-\infty<\ln WTP<+\infty$	20.29	1.63145	17.021	23.578
Model 5					
Truncated Median, Log Transformed		14.54	1.35331	11.817	17.257
Truncated Mean, Log Transformed	$-\infty<\ln WTP<+\infty$	19.61	1.53113	16.531	22.685

Table-A.7.4.2h: Estimated WTP for the NLD Rural Telephone Service in Bangladesh: Bounded Probit Model

	WTP	Std. Error	Confidence Interval	
Model 1				
Median WTP	14.52759	1.01535	12.48717	16.56801
Mean WTP	18.71628	1.36197	15.97930	21.45326
Model 3				
Median WTP	14.44426	1.03320	12.36796	16.52056
Mean WTP	18.35797	1.33656	15.67205	21.04390
Model 5				
Median WTP	14.29308	1.02019	12.24293	16.34322
Mean WTP	17.90737	1.22079	15.45410	20.36064

Table-A.7.4.2i: Estimated WTP for the LDN Telephone Service in Peru: Logit Model

		WTP	Std. Error	Confidence Interval	
Model 1					
Mean WTP	-∝<WTP<+∝	1.4470	0.0362	1.3761	1.5179
Truncated Mean WTP	0<WTP<+∝	1.4625	0.0339	1.3960	1.5290
Truncated Mean WTP	0<WTP< Bmax	1.1414	0.0219	1.0985	1.1843
Model 2					
Mean WTP	-∝<WTP<+∝	1.4477	0.0359	1.3775	1.5180
Truncated Mean WTP	0<WTP<+∝	1.4615	0.0338	1.3952	1.5278
Truncated Mean WTP	0<WTP< Bmax	1.1488	0.0223	1.1051	1.1925
Model 3					
Mean WTP	-∝<WTP<+∝	1.4478	0.0358	1.3776	1.5180
Truncated Mean WTP	0<WTP<+∝	1.4615	0.0338	1.3953	1.5278
Truncated Mean WTP	0<WTP< Bmax	1.1492	0.0223	1.1054	1.1930

Table-A.7.4.2j: Estimated WTP for the LDN Rural Public Telephone Service in Peru: Bounded Probit Model

	WTP	Std. Error	Confidence Interval	
Model 1				
Median WTP	1.382	0.0319	1.3191	1.4442
Mean WTP	1.530	0.0549	1.4227	1.6381
Model 2				
Median WTP	1.382	0.0316	1.3195	1.4436
Mean WTP	1.526	0.0534	1.4215	1.6308
Model 3				
Median WTP	1.381	0.0317	1.3191	1.4433
Mean WTP	1.524	0.0526	1.4205	1.6268

Table-A.7.4.2k: Estimated WTP for International Rural Telephone Service in Bangladesh: Logit Model

		WTP	Std. Error	Confidence Interval	
Model 1					
Mean WTP	-∝<WTP<+∝	66.46313	4.309296	57.80328	75.12299
Truncated Mean WTP	0<WTP<+∝	69.51225	4.28497	60.90127	78.12322
Truncated Mean WTP	0<WTP< Bmax	52.47106	1.584462	49.28696	55.65515
Model 2					
Mean WTP	-∝<WTP<+∝	67.95094	6.083319	55.72606	80.17583

Truncated Mean WTP	0<WTP<+∝	70.0332	5.982912	58.01009	82.05631
Truncated Mean WTP	0<WTP< Bmax	53.68852	1.773058	50.12543	57.25162
Model 3					
Mean WTP	-∝<WTP<+∝	67.75466	6.034249	55.62838	79.88094
Truncated Mean WTP	0<WTP<+∝	69.88081	5.841696	58.14148	81.62013
Truncated Mean WTP	0<WTP< Bmax	53.60218	1.928336	49.72704	57.47731

Table-A.7.4.2l: Estimated WTP for International Rural Telephone Service in Bangladesh: Log Logit Model

		WTP	Std. Error	Confidence Interval	
Model 1					
Truncated Median, Log Transformed		60.24947	4.34499	51.51789	68.98108
Truncated Mean, Log Transformed	-∝<lnWTP<+∝	96.19119	22.26559	51.44684	140.93560
Model 2					
Truncated Median, Log Transformed		61.98497	6.73319	48.45415	75.51583
Truncated Mean, Log Transformed	-∝<lnWTP<+∝	91.10532	21.07497	48.75361	133.45710
Model 3					
Truncated Median, Log Transformed		61.66193	6.57624	48.44647	74.87736
Truncated Mean, Log Transformed	-∝<lnWTP<+∝	90.47755	19.45429	51.38266	129.57240

Table-A.7.4.2m: Estimated WTP for International Rural Telephone Service in Bangladesh: Bounded Probit Model

	WTP	Std. Error	Confidence Interval	
Model 1				
Median WTP	60.90585	4.54528	51.77177	70.03992
Mean WTP	77.12824	7.77511	61.50357	92.75291
Model 2				
Median WTP	61.98541	6.03486	49.85791	74.11291
Mean WTP	75.87428	8.70383	58.38327	93.36528
Model 3				
Median WTP	61.55601	5.93100	49.63722	73.47480
Mean WTP	75.33016	8.26761	58.71578	91.94455

NONPARAMETRIC ESTIMATION

Table-A.7.4.2n: WTP for Local Call in Peru: Turnbull lower bound Nonparametric Mean

bj	cdfj	pdfj	lj	E(WTP)	Std Dev.	Confidence Interval (95%)	
0.00	0.1042	0.1042	0.0000				
0.50	0.5061	0.4019	0.2009				
1.00	0.7605	0.2544	0.2544				
1.50	0.8465	0.0860	0.1290				
2.00	1.0000	0.1535	0.3071	0.8913	0.0075	0.8766	0.9060

Table-A.7.4.2o: WTP for Local Call in Peru: Kriström Nonparametric Mean

bj	cdfj	pdfj	lj	E(WTP)	Std Dev.	Confidence Interval (95%)	
0.00	1.0000	0.0000	0.0000				
0.25	0.8958	0.1042	0.0261				
0.75	0.4939	0.4019	0.3014				
1.25	0.2395	0.2544	0.3180				
1.75	0.1535	0.0860	0.1504				
2.00	0.0000	0.1504	0.3009	1.0968	0.0505	0.9979	1.1957

Table-A.7.4.2p: WTP for LDN Call in Peru: Turnbull Lower bound Nonparametric Mean

bj	cdfj	pdfj	lj	E(WTP)	Std Dev.	Confidence Interval (95%)	
0.0	0.254	0.254	0.000				
1.0	0.577	0.323	0.323				
1.5	0.798	0.221	0.332				
2.0	0.866	0.068	0.136				
2.5	1.000	0.134	0.334	1.125	0.033	1.0596	1.1909

Table-A.7.4.2q: WTP for LDN Call in Peru: Kriström Nonparametric Mean

bj	cdfj	pdfj	lj	E(WTP)	Std Dev.	Confidence Interval (95%)	
0.00	1.000	0.000	0.000				
0.50	0.746	0.254	0.127				
1.25	0.423	0.323	0.404				
1.75	0.202	0.221	0.387				
2.25	0.134	0.068	0.153				
2.50	0.000	0.134	0.334	1.405	0.046	1.3160	1.4947

Table-A.7.4.2r: WTP for Local call in Bangladesh: Kriström Nonparametric Mean

bj	cdfj	pdfj	lj	E(WTP)	Std Dev.	Confidence Interval (95%)	
0.00	1.000	0.000	0.000				
1.70	0.973	0.027	0.023				
7.00	0.704	0.268	1.167				
13.00	0.400	0.304	3.042				
18.00	0.256	0.144	2.226				
23.000	0.000	0.256	5.256	11.715	1.985	7.82532	15.6047

Table-A.7.4.2s: WTP for LDN call in Bangladesh: Kriström Nonparametric Mean

bj	cdfj	pdfj	lj	E(WTP)	Std Dev.	Confidence Interval (95%)	
0.00	1.000	0.000	0.000				
3.40	0.959	0.041	0.070				
16.00	0.471	0.487	4.729				
29.00	0.150	0.321	7.232				
42.00	0.117	0.033	1.176				
42.000	0.000	0.117	5.669	18.875	3.039	12.919	24.831

Table-A.7.4.2t: WTP for International call in Bangladesh: Kriström Nonparametric Mean

bj	cdfj	pdfj	lj	E(WTP)	Std Dev.	Confidence Interval (95%)	
0.00	1.000	0.000	0.000				
26.00	0.833	0.167	2.167				
51.00	0.586	0.248	9.533				
75.00	0.317	0.269	16.950				
100.00	0.321	-0.004	-0.337				
100.00	0.000	0.321	36.058	64.371	14.615	35.725	93.017

Table-A.7.4.2u: WTP for Local call in Bangladesh: Turnbull Lower Bound Nonparametric Mean and Median

bj	cdfj	pdfj	lj	E(WTP)	Std Dev.	Confidence Interval (95%)		Median
1.7	0.027	0.027	0.000					
7	0.296	0.268	0.456					
13	0.600	0.304	2.130					
18	0.744	0.144	1.867					
>18	1.000	0.256	4.615	9.068	2.038382	5.072642	13.063	5.728

Table-A.7.4.2v: WTP for NLD call in Bangladesh: Turnbull Lower Bound Nonparametric Mean and Median

bj	cdfj	pdfj	lj	E(WTP)	Std Dev.	Confidence Interval (95%)		Median
3.4	0.041	0.041	0.000					
16	0.529	0.487	1.657					
29	0.850	0.321	5.143					
42	0.883	0.033	0.960					
>42	1.000	0.117	4.909	12.670	2.351223	8.061358	17.278	14.844

Table-A.7.4.2w: WTP for International call in Bangladesh: Turnbull Lower Bound Nonparametric Mean and Median

bj	cdfj	pdfj	lj	E(WTP)	Std Dev.	Confidence Interval (95%)		Median
26	0.167	0.167	0.000					
51	0.414	0.248	6.438					
75	0.683	0.269	13.721					
100	0.679	-0.004	-0.288					
>100	1.000	0.321	32.051	51.922	13.397	25.664	78.180	59.654

REFERENCES

Akerlof, G.A. 1970 'The Market for Lemons: Qualitative Uncertainty and the Market Mechanism', *Quarterly Journal of Economics*, 84: 488-500.

An, Mark Yuying. 2000. 'A Semiparametric Distribution for Willingness to Pay and Statistical Inference with Dichotomous Choice Contingent Valuation Data', *American Journal of Agricultural Economics*, 82: 487-500.

Ardilla, S. 1993. Guia para la utilizacion de modelos econométricos en aplicaciones del método de valoración contingente. Environmental Division Working Paper ENP-101. Washington DC, Inter-American Development Bank.

Armstrong, M. 2001. 'Access Pricing, Bypass, and Universal Service', *American Economic Review Papers and Proceedings*, 91: 297-301.

Avgerou, C. 1998. 'How Can IT Enable Economic Growth in Developing Countries?', *Information Technology for Development*, 8: 15-28.

Badelt, C. 1985. *Politische Ökonomie der Freiwilligenarbeit: Theoretische Grundlegung und Anwendungen in der Sozialpolitik*. Campus Verlag, Frankfurt/Main.

Bardhan, P. 1989. *The Economic Theory of Agrarian Institutions*. Oxford: Clarendon Press.

Barros, Pedro P. and Seabra, M. C. 1999. 'Universal Service: Does Competition Help or Hurt?', *Information Economics and Policy,* 11: 45-60.

Bateman, I.J. and Willis, K.G. (eds.). 1999. *Valuing Environmental Preferences: Theory and Practice of the Contingent Valuation Method in the US, EC and Developing Countries*, Oxford University Press, London.

Becker, G.S. 1965. 'A Theory of the Allocation of Time', *The Economic Journal*, 75: 493-517.

Bayes, A., von Braun, J. and Akhter, R. 1999. Village Pay Phones and Poverty Reduction: Insights from a Grameen Bank Initiative in Bangladesh, ZEF Discussion Paper 8, Bonn University.

Bayes, A., Hussain, A., Chowdhury, S. (forthcoming). *ICT Services in Bangladesh: Empirical Evidences on Demand and Supply Side.*

Bedi, A. 1999. The Role of ICT in Economic Development – A Partial Survey. ZEF Discussion Paper on Development Policy 7, Bonn University.

Besley, T. Coate, P. and Loury, G. 1993. 'The Economics of Rotating Savings and Credit Associations', *American Economic Review*, 83: 792-810.

Besley, T. and Coate, S. 1995. 'Group Lending, Repayment Incentives and Social *Collateral' Journal of Development Economics*, 46(1): 1-18

Biglaiser, G. 1993. 'Middlemen as Experts', *Rand Journal of Economics*, 24: 212-23.

Binswanger, H.P. and Rosenzweig, M.R. (Eds.) 1984. *Contractual Arrangements, Employment, and Wages in Rural Labour Markets in Asia*. New Haven: Yale University Press.

Blackman, C. 1995. 'Universal service: obligation or opportunity?', *Telecommunications policy*, 19: 171-176.

Bowker, J. and Stoll, J. 1988. 'Use of Dichotomous Choice Nonmarket Methods to Value the Whooping Crane Resource,' *American Journal of Agricultural Economics*, 70: 372-81.

Brook, P.J., and Smith, S.M (eds.). 2001. *Contracting for Public Services*. World Bank and International Finance Corporation, Washington DC.

Cameron, T., and James, M. 1987. 'Efficient Estimation Methods for Use with 'Closed Ended' Contingent Valuation Survey Data,' *Review of Economics and Statistics*, 69: 269-76.

Cannok, G. 2001. 'Expanding Rural Telephony: Output-based Contracts for Pay Phones in Peru', in: Brook, P.J., and Smith, S.M (eds.). 2001. *Contracting for Public Services*. World Bank and International Finance Corporation, Washington DC.

Coase, R.H. 1937. '*The Nature of the Firm*,' *Economica*, 4, 386-405.

Crandall, R., Waverman, L., 2000. *Who pays for universal service? When telephone subsidies become transparent*, Brookings Institution, Washington DC.

Cremer, H., Gasmi, F., Grimaud, A. and Laffont, J-J. 1998. *The Economics of Universal Service: Theory*. The Economic Development Institute of the World Bank.

Crémer, J. 2000. 'The Economics of Internet: Network Externalities and Universal Service Obligation in the Internet,' *European Economic Review*, 44: 1021-1031.

Dasgupta, P. and Stiglitz, JE. 1988, 'Learning-by-doing, market structure, and industrial and trade policies', *Oxford Economic Papers*, 40: 246-268.

de Janvry, A., Fafchamps, M. and Sadoulet, E. 1991. 'Peasant Household Behaviour with Missing Markets: Some Paradoxes Explained,' *The Economic Journal*, 101: 1400-1417.

Demsetz, H. 1968. 'Why Regulate Utilities?', *Journal of Law and Economics*, 11: 55-65.

Duffied, J., and Patterson, D.A. 1991. 'Inference and Optimal Design for a Welfare Measure in Dichotomous Choice Contingent Valuation', *Land Economics*, 67(2): 225-39.

Dunning, J. H. (1980), 'Toward an Eclectic Theory of International Production: Some Empirical Tests', *Journal of International Business Studies*, Spring/Summer 1980, pp.9-31.

Easterly, W. and Levine, R. 1997. 'Africa's Growth Tragedy: Policies and Ethnic Divisions,' *Quarterly Journal of Economics*, 107: 1203-1250.

Evans, W.N., Farrelly, M.C. and Montgomery, E. 1999. 'Do Workplace Bans Reduce Smoking?' *The American Economic Review*, 89: 728-747.

Fahlbeck, Erik. 1996. *Essays in Transaction Cost Economics*. Swedish University of Agricultural Sciences (SLU), Uppsala 1996.

Federal Communications Commission (FCC). 1999. *Connecting The Globe: A Regulator's Guide To Building A Global Information Community*. Washington, DC. http://www.fcc.gov/connectglobe/

Freixas, Xavier and Rochet, Jean-Charles. 1997. *Microeconomics of Banking*. The MIT Press, Cambridge.

Gasmi, F., Laffont, JJ. and Sharkey, WW. 2000. 'Competition, Universal Service Obligation and Telecommunications Policy in Developing Countries', *Information Economics and Policy*, 12: 221-248.

Gertler, P., and Glewwe, P. 1989. The Willingness to Pay for Education in Developing Countries: Evidence from Rural Peru. LSMS Working Paper No. 54. World Bank, Washington DC.

Gertler, P., Locay, L., and Sanderson, W. 1987. 'Are Users Fees Regressive? The Welfare Implications of Health Care Financing in Peru,' *Journal of Econometrics*, 36: 67-88.

Gillet, S.E. 1994. 'Technological Change, Market Structure and Universal Service,' in: Paper presented at the 22nd Annual Telecommunications Policy Research Conference.

Government of Bangladesh (GOB). 'Bangladesh Economic Review 1997, 1998, 1999, 2000.

GOB. Five-Year Plans, 1st, 2nd, 3rd, 4th, 5th.

GOB. Statistical Yearbooks, Bureau of Statistics, Various Issues.

Grameen Bank. http://www.grameen-info.org/bank/supdates.html

Grameen Phone, Annual Reports, 1999, 2000.

Greene, William H. 2000. *Econometric Analysis*. 4th Edition. Prentice-Hall Inc., New Jersey.

Griliches, Z. 1957. 'Hybrid Corn: An Exploration in the Economics of Technological Change,' *Econometrica*, 25(4): 501-522.

Haab, Timothy and McConnell, Kenneth. 1996. Referendum Models and Negative Willingness to Pay: Alternative Solutions, Dept. of Economics, East Carolina University Working Paper No 10.

Haab, Timothy ; McConnell, Kenneth, 1997. A simple Method for Bounding Willingness to Pay Using a Probit or Logit Model, University of Maryland, College Park. Working Paper.

Hanemann, W. M. 1984. 'Welfare Evaluations in Contingent Valuation Experiments with Discrete Responses', *American Journal of Agricultural Economics*, 66: 332-41.

Hanemann, W. M, 1989. 'Welfare Evaluations in Contingent Valuation Experiments with Discrete Response Data: Reply', *American Journal of Agricultural Economics*, 71 (4): 1057-61.

Hanemann, W.M. and Kanninen, B. 1999. 'The Statistical Analysis of Discrete-Response CV Data' in Bateman, I.J. and Willis, K.G. (eds.), *Valuing Environmental Preferences: Theory and Practice of the Contingent Valuation Method in the US, EC and Developing Countries*, Oxford University Press, London.

Hart, O., Shleifer, A., and Vishny, R.W. 1997. 'The Proper Scope of Government: Theory and an Application to Prisons,' *Quarterly Journal of Economics*, 112(4): 1126-61.

Haruvy, E., Prasad, A.1998. Optimal product strategies in the presence of network externalities, *Information economics and policy*, 10: 489-499.

Hausman, J. 1993. *Contingent Valuation: A Critical Assessment*. Amsterdam: North Holland.

Heckman, J. 1979. 'Sample Selection Bias as a Specification Error,' *Econometrica*, 47: 153-161.

Heeks, R. 1999. Information and Communication Technologies, Poverty and Development, Development Informatics Working Paper 5, Institute for Development Policy and Management, University of Manchester.

Hirshleifer, J., and Hirshleifer, D. 1984. *Price Theory and Applications*, 3rd edition. Prentice-Hall, NJ.

Hobbs, Jill E. 1997. 'Measuring the Importance of Transaction Cost in Cattle Marketing,' *American Journal of Agricultural Economics*, 79: 1083-1079.

http://emc-database.com

http://www.grameen-info.org

ITU. 1998. *World Telecommunication Development Report 1998*. International Telecommunication Union (ITU), Geneva.

ITU Data Set 2001. International Telecommunications Unit (ITU), Geneva.

James, E. 1982. Cross-Subsidization by nonprofit organizations: Theory, Evidence and Evaluation, Yale University Program on Nonprofit Organizations Working Paper No. 95.

Jolliffe D. 1998. 'Skills, Schooling, and Household Income in Ghana,' *The World Bank Economic Review*, 12: 81-104.

Jorgenson, D.W. 2001. 'Information Technology and The US Economy,' *American Economic Review*, 91: 1-32.

Just, R., Hueth, D. and Schmitz, A. 1982. *Applied Welfare Economics and Public Policy*. Englewood Cliffs, NJ: Prentice-Hall.

Kaserman, D. L., Mayo, J.W., and Flynn, J. 1990. 'Cross-subsidization in Telecommunications: Beyond the Universal Service Fairy Tale,' *Journal of Regulatory Economics*, 2: 231-249.

Kay, J. 1994. 'Regulating Networks,' In: M.E. Beesley (Ed.), *Regulating Utilities: The Way Forward*. London: Institute of Economic Affairs

Key, N., Sadoulet, E., de Janvry, A. 2000. 'Transaction Costs and Agricultural Household Supply Response,' *American Journal of Agricultural Economics*, 82 (2): 245-269.

Kriström, Bengt. 1990. 'A Nonparametric Approach to Estimation of Welfare Measures in Discrete Response Valuation Studies,' *Land Economics*, 66 (2): 135-139.

Kubicek, H., Dutton, W., and Williams, R. (Eds.). 1997. *The Social Shaping of Information Superhighways: European and American Roads to the*

Information Society. Frankfurt: Campus Verlag and New York: St. Martin Press.

LaFrance, J.T. 1991. 'Consumer's Surplus Versus Compensating Variation Revisited,' *American Journal of Agricultural Economics*, 73(5): 1496-1507.

Lal, D. 1997. 'Private Provision of Public Goods and Services.' In Herbert Giersch (Ed.), *Privatization at the End of the Century*, Springer: Berlin. 329-361.

Leff, NH. 1984. 'Externalities, Information costs, and Social Benefit-Cost Analysis for Economic Development: An example from Telecommunications,' *Economic Development and Cultural Change*, 32: 255-276.

Li, Yiting. 1998. 'Middlemen and Private Information,' *Journal of Monetary Economics*, 42: 131-159.

Maddala, G.S. 1983. *Limited-Dependent and Qualitative Variables in Econometrics*. Cambridge University Press, New York.

Management Consulting Group (MCG). 2001. *Survey on ICTs*, Dhaka, Bangaldesh.

Mansell, R. 1999. 'Information and Communication Technologies for Development: Assessing the Potential and the Risks,' *Telecommunications Policy*, 23: 35-50.

Maral, G. and M. Bousquet. 1998. *Satellite Communications Systems: Systems, Techniques and Technology*. 3rd Edition, John Wiley & Sons Ltd. England.

Masten, S.E., Meehan, J.W., and Snyder, E.A. 1991. 'The Costs of Organization,' *Journal of Law, Economics and Organization*, 7: 1-25.

McConnell, K.E. 1995. Issues in Estimating Benefits with Non-Market Methods, Office of the Chief Economist Working Paper Series 308. Inter-American Development Bank. Washington DC.

McFadden, Daniel. 1994. 'Contingent Valuation and Social Choice,' *American Journal of Agricultural Economics*, 76: 689-708.

Messonnier, M.L., Bergstrom, J.C., Cornwell, C.M., Teasley, R.J., and Cordell, H.K. 2000. 'Survey Response-Related Biases in Contingent Valuation: Concepts, Remedies, and Empirical Application to Valuing Aquatic Plant Management,' *American Journal of Agricultural Economics*, 83: 438-450.

Minten, B. and Kyle, S. 1999. 'The Effect of Distance and Road Quality on Food Collection, Marketing Margins and Traders' Wage: Evidence from the former Zaire,' *Journal of Development Economics*, 60: 467-495.

Mitchel, R.C. and Carson, R.T. 1989. *Using Surveys to Value Public Goods: The Contingent Valuation Method*. Washington DC: Resources for the Future.

Mitchell, B.M. 1978. 'Optimal Pricing of Local Telephone Service', *American Economic Review*, 68 (4): 517-537.

Morales-Gomez, D. Melesse, M. 1998. 'Utilizing Information and Communication Technologies for Development: The Social Dimensions,' *Information Technology for Development* 8: 3-13.

Mueller, M. 1993. 'Universal Service in Telephone History: A Reconstruction,' *Telecommunications Policy*, 17: 352-369.

OECD. 1988. *New Technologies in the 1990s: A Socio-Economic Strategy*, Paris.

Okiniwa Charter on Global Information Society. 2000. http://www.dotforce.org/reports/it1.html

OSIPTEL-FITEL. 1999. '*El acceso universal y la politica del FITEL*'. Serie: Estudios en Telecommunicaciones No. 5. Lima.

Otsuka, K., Chuma, H. and Hayami, Y. 1992. 'Land and Labor Contracts in Agrarian Economies: Theories and Facts,' *Journal of Economic Literature*, 30: 1965-2018.

Panzar, John C. 2000. 'A Methodology for Measuring the Costs of Universal Service Obligations,' *Information Economics and Policy*, 12: 211-220.

Richardson, D., Ramirez, R., and Haq, M. 2000. '*Grameen Telecom's Village Phone Programme in Rural Bangladesh: a Multi-Media Case Study*,' TeleCommons Development Group (TDG), Ontario.

Robinson, P.M. 1988. 'Root-N-Consistent Semiparametric Regression,' *Econometrica*, 56: 931-954.

Röller, L-H and Waverman, L. 2001. 'Telecommunications Infrastructure and Economic Growth: A Simultaneous Approach,' *The American Economic Review*, 4: 909-923

Romer, P. 1986. 'Increasing Returns and Long-Run Growth,' *Journal of Political Economy* 94: 1002-37.

_________. 1990. 'Endogenous Technological Change,' *Journal of Political Economy* 98: 71-102.

Rose-Ackerman, S. 1996. 'Altruism, Nonprofits, and Economic Theory,' *Journal of Economic Literature*, 34: 701-728.

Roy K. Ziemek S. 2000. On the Economics of Volunteering, ZEF-Discussion Paper 31, Bonn University.

Rubinstein, A. and A. Wolinsky. 1987. 'Middlemen,' *Quarterly Journal of Economics*, 102: 581-593

Sadoulet, E., and de Janvry, A. 1995. *Quantitative Development Policy Analysis*. Baltimore: The Johns Hopkins University Press.

Salamon, LM. and Anheier, HK. 1998. *The Emerging Sector: An Overview*. The John Hopkins University, Baltimore.

Small, K.A., and Rosen, H.S. 1981. 'Applied Welfare Economics with Discrete Classic Models,' *Econometrica*, 49 (1): 105-130.

Sorana, V. 2000. 'Auctions for Universal Service Subsidies,' *Journal of Regulatory Economics* 18: 33-58.

Stiglitz, J.E. 1974. 'Incentives and Risk Sharing in Sharecropping', *Review of Economic Studies*, 41: 219-255.

Stiglitz, J.E. 1988. 'Economic Organization, Information, and Development', in: H. Chenery and T.N. Srinivasan, (eds.), '*Handbook of Development Economics' Volume 1: 92-160.* Elsevier Science Publishers.
Stiglitz J.E. 1990. 'Peer Monitoring and Credit Markets,' *The World Bank Economic Review*, 4: 351-366.
Strauss, J. 1984. 'Joint Determination of Food Consumption and Production in Rural Sierra Leone Estimates of a Household-Firm Model,' *Journal of Development Economics*, 14: 77-103.
Sussman, G. 1997. *Communication, Technology, and Politics in the Information Age*. London: Sage.
Tapscott, D. 1995. *The Digital Economy: Promise and Peril in the Age of Networked Intelligence*. New York: McGraw-Hill.
The Financial Express, January 15, 2001.
Tobin, J. 1958. 'Estimation of Relationship for Limited Dependent Variables,' *Econometrica*, 26: 24-36.
Torero, M. 2000. The Access and Welfare Impacts of Telecommunications Technology in Peru, ZEF Discussion Paper 27, Bonn University.
Torero, M., Schroth, E., and Pasco Font, A. 2001. Are Telephone Consumers Better after the Privatization of Telecommunications in Peru. Paper presented at GDN Conference 2001 in Rio de Janeiro.
Townsend, Robert M. 1978. 'Intermediation with Costly Bilateral Exchange,' *Review of Economic Studies*, 45: 417-45.
UNDP. 2001. Human Development Report 2001, Geneva.
Valletti TM. 2000. 'Introduction: Symposium on Universal Service Obligation and Competition,' *Information Economics and Policy*, 12: 205-210.
van de Ven, W.P.M.M. and van Praag, B.M.S. 1981. 'The Demand for Deductibles in Private Health Insurance: A Probit Model with Sample Selection,' *Journal of Econometrics*, 17: 229-252.
Vaughan, W.J., Russell, C.S., Rodriguez, D.J., and Darling, A.H. 1999. Willingness to Pay: Referendum Contingent Valuation and Uncertain Benefits. Sustainable Development Department Technical Papers Series, Inter-American Development Bank, Washington DC.
Wellenius, B., Braga, C.A.P., Qiang, C.Z-W. 2000. 'Investment and Growth of the Information Infrastructure: Summary Results of a Global Survey,' *Telecommunications Policy*, 24: 639-643.
Weller, D. 1999. 'Auctions for Universal Service Obligations,' *Telecommunications Policy*, 23: 645-674.
Williamson, O.E. 1975. *Markets and Hierarchies: Analysis and Antitrust Implications*. The Free Press, New York.
___________. 1991. Comparative Economic Organization: The Analysis of Discrete Structural Alternatives. In: Williamson, O.E and Masten S.E (eds.) 1995. *Transaction Costs Economics*, Vol. 1: Theory and Concepts. Aldershot: Edward Elgar Publishing.
World Bank. 1997. Environment Matters, Winter/Spring, Washington DC.

World Bank. 2001. World Development Indicators 2001. Washington DC.
World Bank and Bangladesh Center for Advanced Studies (BCAS), 1998. Bangladesh 2020: A Long-Run Perspective Study. http://worldbank-bangladesh.org

Development Economics and Policy

Series edited by Franz Heidhues and Joachim von Braun

Band 1 Andrea Fadani: Agricultural Price Policy and Export and Food Production in Cameroon. A Farming Systems Analysis of Pricing Policies. The Case of Coffee-Based Farming Systems. 1999.

Band 2 Heike Michelsen: Auswirkungen der Währungsunion auf den Strukturanpassungsprozeß der Länder der afrikanischen Franc-Zone. 1995.

Band 3 Stephan Bea: Direktinvestitionen in Entwicklungsländern. Auswirkungen von Stabilisierungsmaßnahmen und Strukturreformen in Mexiko. 1995.

Band 4 Franz Heidhues / François Kamajou: Agricultural Policy Analysis – Proceedings of an International Seminar, held at the University of Dschang, Cameroon on May 26 and 27 1994, funded by the European Union under the Science and Technology Program (STD). 1996.

Band 5 Elke M. Förster: Protection or Liberalization? A Policy Analysis of the Korean Beef Sector. 1996.

Band 6 Gertrud Schrieder: The Role of Rural Finance for Food Security of the Poor in Cameroon. 1996.

Band 7 Nestor R. Ahoyo Adjovi: Economie des Systèmes de Production intégrant la Culture de Riz au Sud du Bénin: Potentialités, Contraintes et Perspectives. 1996.

Band 8 Jenny Müller: Income Distribution in the Agricultural Sector of Thailand. Empirical Analysis and Policy Options. 1996.

Band 9 Michael Brüntrup: Agricultural Price Policy and its Impact on Production, Income, Employment and the Adoption of Innovations. A Farming Systems Based Analysis of Cotton Policy in Northern Benin. 1997.

Band 10 Justin Bomda: Déterminants de l'Epargne et du Crédit, et leurs Implications pour le Développement du Système Financier Rural au Cameroun. 1998.

Band 11 John M. Msuya: Nutrition Improvement Projects in Tanzania: Implementation, Determinants of Performance, and Policy Implications. 1998.

Band 12 Andreas Neef: Auswirkungen von Bodenrechtswandel auf Ressourcennutzung und wirtschaftliches Verhalten von Kleinbauern in Niger und Benin. 1999.

Band 13 Susanna Wolf (ed.): The Future of EU-ACP Relations. 1999.

Band 14 Franz Heidhues / Gertrud Schrieder (eds.): Romania – Rural Finance in Transition Economies. 2000.

Band 15 Katinka Weinberger: Women's Participation. An Economic Analysis in Rural Chad and Pakistan. 2000.

Band 16 Christof Batzlen: Migration and Economic Development. Remittances and Investments in South Asia: A Case Study of Pakistan. 2000.

Band 17 Matin Qaim: Potential Impacts of Crop Biotechnology in Developing Countries. 2000.

Band 18 Jean Senahoun: Programmes d'ajustement structurel, sécurité alimentaire et durabilité agricole. Une approche d'analyse intégrée, appliquée au Bénin. 2001.

Band 19 Torsten Feldbrügge: Economics of Emergency Relief Management in Developing Countries. With Case Studies on Food Relief in Angola and Mozambique. 2001.

Band 20 Claudia Ringler: Optimal Allocation and Use of Water Resources in the Mekong River Basin: Multi-Country and Intersectoral Analyses. 2001.

Band 21 Arnim Kuhn: Handelskosten und regionale (Des-)Integration. Russlands Agrarmärkte in der Transformation. 2001.

Band 22 Ortrun Anne Gronski: Stock Markets and Economic Growth. Evidence from South Africa. 2001.

Band 23 Patrick Webb / Katinka Weinberger (eds.): Women Farmers. Enhancing Rights, Recognition and Productivity. 2001.

Band 24 Mingzhi Sheng: Lebensmittelkonsum und -konsumtrends in China. Eine empirische Analyse auf der Basis ökonometrischer Nachfragemodelle. 2002.

Band 25 Maria Iskandarani: Economics of Household Water Security in Jordan. 2002.

Band 26 Romeo Bertolini: Telecommunication Services in Sub-Saharan Africa. An Analysis of Access and Use in the Southern Volta Region in Ghana. 2002.

Band 27 Dietrich Müller-Falcke: Use and Impact of Information and Communication Technologies in Developing Countries' Small Businesses. Evidence from Indian Small Scale Industry. 2002.

Band 28 Wolfram Erhardt: Financial Markets for Small Enterprises in Urban and Rural Northern Thailand. Empirical Analysis on the Demand for and Supply of Financial Services, with Particular Emphasis on the Determinants of Credit Access and Borrower Transaction Costs. 2002.

Band 29 Wensheng Wang: The Impact of Information and Communication Technologies on Farm Households in China. 2002.

Band 30 Shyamal K. Chowdhury: Institutional and Welfare Aspects of the Provision and Use of Information and Communication Technologies in the Rural Areas of Bangladesh and Peru. 2002.